A Relatively Painless Guide to Special Relativity

A Relatively Painless Guide to Special Relativity

DAVE GOLDBERG

The University of Chicago Press
Chicago and London

The University of Chicago Press, Chicago 60637
The University of Chicago Press, Ltd., London

Published 2023

Printed in the United States of America

32 31 30 29 28 27 26 25 24 23   1 2 3 4 5

ISBN-13: 978-0-226-82542-7 (cloth)
ISBN-13: 978-0-226-82185-6 (paper)
ISBN-13: 978-0-226-82186-3 (e-book)

DOI: https://doi.org/10.7208/chicago/9780226821863.001.0001

Library of Congress Cataloging-in-Publication Data

Names: Goldberg, Dave, 1974–, author.
Title: A relatively painless guide to special relativity / Dave Goldberg.
Description: Chicago; London: The University of Chicago Press, 2023. |
    Includes bibliographical references and index.
Identifiers: LCCN 2022050654 | ISBN 9780226825427 (cloth) |
    ISBN 9780226821856 (paperback) | ISBN 9780226821863 (ebook)
Subjects: LCSH: Special relativity (Physics)—Textbooks. |
    LCGFT: Textbooks.
Classification: LCC QC173.65 .G64 2023 | DDC 530.11—dc23/eng20221121
LC record available at https://lccn.loc.gov/2022050654

♾ This paper meets the requirements of ANSI/NISO Z39.48-1992
(Permanence of Paper).

Contents

Preface

We're imbued with an intuitive grasp of the physical universe. Play a game of catch with a friend, and with a little practice, you can adjust the direction and speed so that the ball reaches them almost every time. We run and jump and play airplane with the small people in our lives, spinning them around without ever consciously doing a calculation about angular momentum, calculating centripetal force, or even needing to convert to rotating coordinates.

This is all well and good at normal terrestrial speeds, but most of the interesting stuff—the world of particle accelerators and supernova explosions—does *not* happen at human speeds, and the physics of those high speeds is much, much different.

In this book you'll learn about how space and time *work*. We'll discuss how a physical theory can be developed and how familiar, albeit complicated, ideas like Maxwell's equations of electromagnetism can be developed simply and seemingly from first principles. You'll learn about special relativity itself because it's amazing, but also because it's the foundation for modern particle physics and general relativity and so much else. We'll be developing rules for writing down the laws of the universe and figuring out what's possible and what's not. Relativity, and the math surrounding it, will form the basis for not only this work but also your future study in the wider world of advanced physics.

Finally, a note of welcome to lifelong learners: Some of you may have read qualitative or popular science versions of relativity and wanted to delve deeper to get a better insight into the weirdness of time dilation or the ephemeral nature of matter. Maybe you have a mathematical itch that needs scratching. Since I can't anticipate every possible course sequence for even more traditional learners, I generally only make assumptions about mathematical preparation and put in reminders (or first introductions, as the case may be) before presenting any new physical concepts. I've included answers to odd problems at the end of the book so both types of readers can tell if they're on the right track.

A Note to Instructors

Special relativity is often treated like an extra goody in undergraduate physics education. It's something that students are either expected to pick up in a few random lectures in a mechanics or electromagnetism course, or else it's presented as a combination of geometric and logical puzzles—seemingly with the premise of getting the novice student to conclude that the universe is weird and puzzling. But special relativity is so fundamental to modern physics— it's the canvas on which electromagnetism, particle physics, field theory, and ultimately general relativity are based—that getting a relativistic intuition isn't just a luxury; it's a requirement.

This book was the product of a course I taught during the 2020–2021 pandemic. Because of the limitations of remote lecturing, I needed to make sure my notes were clear (since a student's connection could drop out at any time), and as I was adjusting on the fly, I quickly diverged from the established text for the course, the estimable *Spacetime Physics* by Edwin Taylor and John Wheeler [41], the standard for undergraduate relativity. This text is written at a similar (or slightly more advanced) level, but with a much different focus. It is aimed at the sophomore or junior physics major who is acquainted with introductory calculus; has perhaps encountered divergence, gradient, and curl in Cartesian coordinates; and has had an introductory mechanics and electromagnetism sequence that presented Maxwell's equations in some form. In short, it's aimed at students who are just getting ready to enter a more sophisticated world of symmetry and unification—but before they've been conditioned to think that classical mechanics or electromagnetism is really just a long list of equations and solved special cases.

I originally taught this material in about 25 hours of lecture, suggesting that for most semester institutions, this might be combined as a half-term three-credit course or (depending on the selection of chapters) as a two- or three-credit quarter-term course. Later sections, including those on electromagnetism and introductory general relativity, may be cut for time (though they are extremely fun topics, and I wouldn't drop them if you can help it). In addition, some sections are denoted with an asterisk. They can be skipped, depending on the level and interest of the students.

Finally, a word on what this book is *not*. It is not a history book, or a biography of Albert Einstein, George FitzGerald, Hermann Minkowski, or anyone else. This is not primarily about personalities or the discovery itself, though I include a half dozen books for further reading at the end of each chapter, and especially early on, many of those are particularly readable histories of science. I cite the original papers and give credit where credit is due, but one of the beautiful things about relativity is how inevitable it seems once the postulates are spelled out—and how much can be directly inferred from those principles. The goal of this work is to train an intermediate student to anticipate how a physical theory can be constructed. Physicists, as you know, will often describe a theory as "elegant," as though the concept is intuitive. There can be no better training ground for developing that intuition than relativity.

Space

Euclid of Alexandria circa 300 BCE. His *Elements* provided much of the motivation for Einstein's early mathematical education, and his reasoning in space will form the foundation for much of our work in spacetime. Woodcut by André Thevet, 1584.

1.1 What Relativity Is

Even outside of physics circles, one equation has become so famous that it's transcended the discipline:

$$E = mc^2.$$

This equation forms the basis for the power of fusion in the Sun, for nuclear weapons, and for the creation and annihilation of matter and antimatter in the early universe. It also made Einstein an international celebrity. We won't start off by talking about mass and energy, but instead, we need to think about perspective. The *relative* in *relativity* comes from the challenge of converting one perspective to another.

At its heart, relativity is fundamentally about the study of measurement: how we measure the distance between two points in space, and ultimately, two points in space and time. In ordinary space, measuring distances seems straightforward: mark off two points on the ground and run a tape measure from one to the other.

Like many tasks in the mathematical sciences, it's far easier to do a thing than to describe it with equations. The vibrations of a guitar string require a knowledge of Newton's laws, tension, and differential equations to describe accurately, but even a little kid will know what happens if you pluck it. The same is true here. It's easy to *measure* distances but somewhat trickier to describe what we *mean* by measuring distances. Fortunately, we have the benefit of physical intuition. We'll know when we're done whether our mathematics makes sense. (That same intuition will break down spectacularly shortly after we introduce the mechanics of spacetime.) With that in mind, we can make a concise statement about what relativity is. [7, 13]

Relativity is the idea that (within certain constraints) it does not matter who is making the measurement, how they are oriented, or how or whether they are moving, some measurements will be the same no matter how you look at them.

But Einstein himself urged caution [43]:

> The meaning of relativity has been widely misunderstood. Philosophers play with the word, like a child with a doll. Relativity, as I see it, merely denotes that certain physical and mechanical facts, which have been regarded as positive and permanent, are relative with regard to certain other facts in the sphere of physics and mechanics. It does not mean that everything in life is relative and that we have the right to turn the whole world mischievously topsy-turvy.

It will turn out that, to a very significant degree, the study of relativity will be very much the study of symmetry. As the mathematician Hermann Weyl [45] put it:

> A thing is symmetrical if there is something you can do to it so that after you have finished doing it, it looks the same as before.

Our central goal in relativity will be to show that we can stretch and rotate space and time and that some *thing* will remain as it was before. We'll spend the first couple of chapters figuring out what that *thing* is.

1.2 Measuring Distances

When Einstein was a boy, one of his first inspirations was Euclid's *Elements* [21], a book written around 300 BCE in which the author set out a few simple rules and from them—using only a straight edge and a compass—was able to derive almost all of the fundamental rules of geometry. As Einstein [6] put it:

> Here were assertions … which—though by no means evident—could nevertheless be proved with such certainty that any doubt appeared to be out of the question. This lucidity and certainty made an indescribable impression upon me.

Or, more briefly:

> If Euclid failed to kindle your youthful enthusiasm, then you were not born to be a scientific thinker.

Starting with seemingly* self-evident assumptions like "a straight line segment can be drawn joining any two points" and "given any straight line segment, a circle can be drawn having the segment as radius and one endpoint as the center," Euclid proved a number of geometric theorems, including Pythagoras's:

$$A^2 + B^2 = C^2$$

for a right triangle, as shown in figure 1.1.

The Pythagorean theorem is one of the most useful relations in all of geometry, because it allows us to measure distances. If you're in a gridded city (with uniform, square blocks) and travel A streets east and B streets north, then you've covered a total distance of C as the crow flies.

1.2.1 TENSORS

Just as Euclid limited his readers to using only a few simple tools in order to understand the geometry of the universe, I'm going to limit us to one: rulers. Almost everything we're going to do will be predicated on figuring out the distances between nearby points.

---

*The mathematicians of the nineteenth century showed that *seemingly* is the operative word. Space can be curved, and that changes the rules of geometry considerably. Fortunately for us, however, this book will be focused on the flat space of special relativity. Indeed, flat space is what makes special relativity so special.

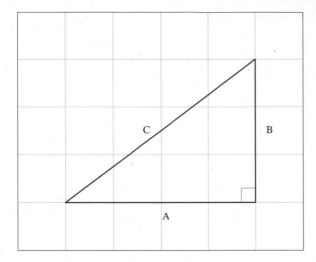

F I G U R E 1.1. The classic Pythagorean triple 3-4-5 right triangle

Let's begin with a fairly typical notation for writing down the position of a particle:

$$\vec{r} = x\hat{i} + y\hat{j} + z\hat{k},$$

where $\hat{i}$ is the unit vector in the x-direction,* $\hat{j}$ the unit vector in the y-direction, and so on. Give the ordered triple $\langle x, y, z \rangle$, and we can identify a point in space.

For most calculations, we can skip writing the unit vectors entirely and simply write down the components themselves:

$$x^i = \begin{pmatrix} x \\ y \\ z \end{pmatrix}. \tag{1.1}$$

This notation might be a little unfamiliar to you. The i in x^i isn't an exponent; it's an index. It's a way of writing out all possible elements of an ordered triple ($i = 1, 2$ or 3). Thus x^2, for instance, is what we'd more commonly call y.

The position vector itself could be written out as

$$\vec{r} = \sum_i x^i \vec{e}_i, \tag{1.2}$$

---

*Different texts use different notation. Some use $\hat{x}$ for the unit vector in the x-direction. Some use **i**. Some use $\vec{e}_1$, as we will when writing out unit vectors in generality. We'll be using the $\hat{i}$ notation for the simple reason that it's easier for instructors to write unambiguously on a whiteboard.

where $\vec{e}_i$ is a generic way of writing the set of three different unit vectors. In Euclidean space,

$$\vec{e}_i = \begin{pmatrix} \hat{i} & \hat{j} & \hat{k} \end{pmatrix}.$$

We happen to live in a three-dimensional universe,* which means that a position vector (or other vectors representing physical quantities) is going to typically have three numbers. Going forward, we're going to refer to "ordinary" vectors, the kind with arrows over them, as **3-vectors**. What we're building up here is known as **tensor** notation. When you first encounter them (now?), tensors are going to appear to be a bunch of odd-looking objects:

$$x^i, \ g_{ij}, \ T^{\mu\nu}.$$

We'll encounter all of these in this book. A tensor is really just a list of numbers. For each index (in three-dimensional space), you have three possible values—1, 2, 3—and each slot can have a different number.

A tensor with two indices (known as a **rank-2 tensor**) would have nine possible values, one for every possible permutation of the first and second index. You may think of it as a 3×3 array. Rank-2 tensors frequently show up in fluid and continuum mechanics, for instance. They might describe the flow of gas molecules at a particular point in space and time. A simple vector (the bulk flow) wouldn't be up to the challenge. To completely describe the velocities in the gas, you'd also want to know the pressure and the vorticity, which can be contained within a rank-2 tensor. There are other tensors as well, including an angular momentum tensor, which encodes $\vec{x} \times \vec{p}$ within a collection of particles, or the stress and strain tensors, which describe the forces and distortions for extended continua.

For now, though, we'll just content ourselves with the knowledge that we *can* collect numbers in whatever rank tensor we like.

Students new to tensors might suppose that the choice to write indices "upstairs"[†] for vector components or "downstairs"[‡] for unit vectors is completely arbitrary or even irrelevant. For now, it's just a "because I told you so" rule, but we'll justify this later on; the choice, it turns out, matters a great deal. As we develop tensor formalism, I'm going to start by giving you a list of rules, and while they may seem a little pedantic, hopefully the notation will look sensible as we're doing it.

---

*Everything we're doing generalizes quite nicely to four or five dimensions, or to one or two. Nevertheless, the term *3-vector* will be used synonymously with space-vector.

[†]Contravariant, if you're looking to be more technical.

[‡]Covariant.

Example: Making a Rank-2 Tensor

Consider two vectors:

$$u^i = \begin{pmatrix} 3 \\ 5 \\ -1 \end{pmatrix}; \quad v^j = \begin{pmatrix} 2 \\ 0 \\ 2 \end{pmatrix}.$$

What is $A^{ij} = u^i v^j$?

Solution

This is what is known as an **outer product**. Which index is which matters. Multiplying the first few terms out by hand, we get $A^{11} = u^1 v^1 = 6$, $A^{12} = u^1 v^2 = 0$, and so on.

We could list the elements of the tensor by hand (e.g., $A^{13} = 6\ldots$), but it's clearly tidier to collect everything in a matrix-style block. Tidier, but a little ambiguous. We need to specify that the first index gives the row and the second the column, and then we can write

$$A^{ij} = \begin{pmatrix} 6 & 0 & 6 \\ 10 & 0 & 10 \\ -2 & 0 & -2 \end{pmatrix}.$$

But we could have chosen the opposite convention. And if we'd had a rank-3 tensor or higher? Good luck trying to write *that* as a matrix!

Tensor math will comprise a set of rules to deal with not only vectors but also more complicated sets of numbers, and, if we follow a few straightforward rules, we'll find ourselves utterly incapable of writing down equations that *aren't* consistent with the principles of relativity.*

1. What you label an index doesn't really matter.
    By themselves, x^i and x^j mean the same thing: the components of a position 3-vector.

That is, x^i, x^j, and x^k are each a list of three numbers, and, more importantly, the *same* set of three numbers. For those of you who program computers, this is just the arbitrary variable that you include on the inside of a loop. What

*I've put a complete set of the relevant tensor rules and notation in Section A.2 of the Appendix at the end of the book.

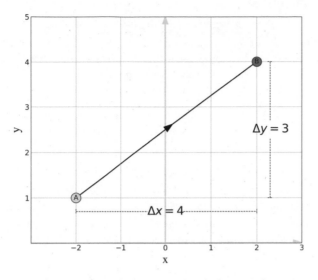

FIGURE 1.2. A two-dimensional particle trajectory with a displacement that conveniently makes a Pythagorean triple

you call it doesn't matter, but by convention, we're going to use Roman indices to indicate three spatial dimensions. Greek, as we'll see, will be reserved for spacetime.

With that in mind, let's consider a trajectory in the xy-plane (illustrated in figure 1.2):

$$x_A^i = \begin{pmatrix} -2 \\ 1 \\ 0 \end{pmatrix}; \quad x_B^i = \begin{pmatrix} 2 \\ 4 \\ 0 \end{pmatrix}.$$

We want to use our measuring tool to figure out the total distance between x_A^i and x_B^i. We have a mathematical approach to doing that: the Pythagorean theorem,

$$(\text{distance})^2 = \Delta x^2 + \Delta y^2,$$

where Δx and Δy are the components of the **displacements**. We can compute the two of them at once using our first tensor equation,

$$\Delta x^i = x_B^i - x_A^i = \begin{pmatrix} 4 \\ 3 \\ 0 \end{pmatrix},$$

which is simply three equations merged into one. Implicit in this equation is the second tensor rule:

2. Equations need to have matching indices—and that includes the location of the index.

    You'll note that the equation above has three terms (one on the left and two on the right), and all of them have an i "upstairs." Thus, there's an unambiguous relation, $\Delta x = x = x_B - x_A$, with similar relations for y and z.

In this case, $(distance)^2$ is 25, so $d = 5$, using the Pythagorean theorem. But we can write the same thing in the form of a tensor equation—albeit one that requires one more rule and the introduction of one special tensor:

3. The **Einstein Summation Convention.**

    First, the special rule. Scalars are some of the most useful tools in physics, which is why it is often convenient to use energy rather than momentum to do calculations. In particular, we often want to combine two vectors into a scalar using a **dot product**. This is precisely why the work-kinetic energy theory ($W = \vec{F} \cdot \Delta \vec{r}$) is such a handy equation.

    To produce scalars using tensor math, there's a nice, tidy rule that tells you what to do if you have the same index upstairs and downstairs on the same side of an equation. That is, if you have a vector, A^i, and another tensor, B_i (one downstairs index is known as a **1-form**), then

$$A^i B_i = A^1 B_1 + A^2 B_2 + A^3 B_3. \tag{1.3}$$

Just add them all up! It looks a lot like a dot product, no? Just a couple of quick caveats. First, it only works if the indices match. You can't do the summation if you have $A^i B_k$, for instance. Second, it only works if one index is upstairs and the other is downstairs. $A^i C^i$, for instance, isn't a valid tensor term.

Example: Summation

Consider the components of a vector, $A^i = \begin{pmatrix} 4 \\ 0 \\ -2 \end{pmatrix}$, and a 1-form,*

$B_k = (2 \; 5 \; 3)$. What is $A^i B_i$?

Solution

Recalling that the index in the definition is arbitrary, the sum is computed via

$$A^i B_i = (4)(2) + (0)(5) + (-2)(3) = \boxed{2}.$$

$A^i B_i$ is a scalar.

---

*One stylistic convention that I use for my own clarity is to write the components of vectors (with an upstairs index) as a column vector, but 1-forms (with a downstairs index) as a row vector. As both are simply lists of numbers, it doesn't matter, but I find it prevents me from mixing them up.

1.2.2 THE METRIC

Now for the new, special tensor—one that's going to be central to our measurements in relativity. It's important enough that, all by itself, it constitutes a tensor rule:

4. The **metric** is a rank-2 tensor that facilitates taking dot products and raising and lowering indices.

Rather than spell it out in words, a common example will probably help. Suppose two points are separated by a displacement, dx^i. In that case, we can compute the distance (squared) between them via a general relationship:

$$dl^2 = g_{ij}dx^i dx^j, \qquad (1.4)$$

where the metric itself is that g object. In Cartesian coordinates, it has the values

$$g_{ij} = \begin{pmatrix} 1 & 0 & 0 \\ 0 & 1 & 0 \\ 0 & 0 & 1 \end{pmatrix}. \qquad (1.5)$$

This looks really, really complicated, but it turns out to be much simpler if we write it out:

$$dl^2 = g_{ij}dx^i dx^j$$
$$= \sum_{i=1}^{3}\sum_{j=1}^{3} g_{ij}dx^i dx^j$$
$$= g_{11}dx^1 dx^1 + g_{12}dx^1 dx^2 + g_{13}dx^1 dx^3 + \cdots + g_{33}dx^3 dx^3$$
$$= (1)dx^1 dx^1 + (0)dx^1 dx^2 + (0)dx^1 dx^3 + \cdots + (1)dx^3 dx^3$$
$$= (dx)^2 + (dy)^2 + (dz)^2.$$

In general, I won't be *quite* so pedantic about multiplying out the metric, since it's diagonal. Only dx^2, dy^2, and dz^2 terms will be included, so if you're used to it, you can figure out what a particular metric means by inspection. And for what it's worth, we'll only be using this and one other metric in this book. More generally, the metric allows you to take a dot product with any two vectors:

$$\vec{A} \cdot \vec{B} = g_{ij} A^i B^j = |\vec{A}||\vec{B}| \cos\phi, \tag{1.6}$$

where ϕ is the angle between the vectors. If you pay attention to the index accounting, all of the upstairs and downstairs indices "contract" with one another, so we aren't left with any. A dot product of two vectors gives you a scalar.

I've been referring to these particular components of g_{ij} as *the* metric, but in actuality, it's really *a* metric. There are lots of them. In two-dimensional polar space (r, θ), the metric is a bit different:*

$$g_{ij}^{(polar)} = \begin{pmatrix} 1 & 0 \\ 0 & r^2 \end{pmatrix}.$$

This is just another way of expressing a two-dimensional flat surface; it's our familiar Euclidian metric (eq. 1.5) expressed in different coordinates. We could try writing down the equations in special relativity in polar coordinates, but it'll get ugly—in large part because the metric is more complicated.

Here's a third metric, one describing a unit sphere using the polar angles (θ, ϕ):

$$g_{ij}^{(sphere)} = \begin{pmatrix} 1 & 0 \\ 0 & \sin^2\theta \end{pmatrix}.$$

This cue ball metric describes a fundamentally curved surface—which means that it's beyond the scope of this book. Should you continue on to general relativity, you're going to spend a lot of time thinking about curved surfaces.

All of this is to say that in this text we will confine ourselves almost exclusively to metrics like equation 1.5 that are diagonal and constant in space and time.

---

*In one of the optional sections at the end of the chapter, we'll derive this directly.

Example: Dot Products

Consider two vectors, $u^i = \begin{pmatrix} 2 \\ 3 \\ 0 \end{pmatrix}$ and $v^i = \begin{pmatrix} 5 \\ -4 \\ 1 \end{pmatrix}$. What is $\vec{u} \cdot \vec{v}$?

Solution

Using the metric notation, the dot product may be computed as

$$\vec{u} \cdot \vec{v} = g_{ij} u^i v^j$$

$$= g_{11} u^1 v^1 + g_{22} u^2 v^2 + g_{33} u^3 v^3$$

$$= (2)(5) + (3)(-4) + (0)(1)$$

$$= \boxed{-2},$$

where in the second line, we simply used the fact that only the diagonal elements of the metric are nonzero.

It may seem like we've done a lot of work for very little payoff—the final result is just the ordinary dot product that you've been doing for years. But be patient. All of this pedantry will pay off in due course.

1.3 Coordinate Transformations

1.3.1 FRAMES OF REFERENCE

Many dynamics problems begin with placing a particle at such-and-such a place with this-or-that velocity, and asking us to predict its evolution. All of which assumes we share the same perspective.

When talking about a point in space (or, ultimately, spacetime), we designate a position. The object itself may be a ball or an atom or a person. It doesn't really matter. All that matters is that we have a specific set of coordinates in a particular **reference frame**.

You know the old thought experiment where you're meant to imagine a trillion monkeys on a trillion typewriters working for a trillion years? Those monkeys, we're told, will ultimately churn out the works of Shakespeare. A frame of reference is similar, if a little more mundane: Imagine a trillion physicists with a trillion identical rulers and a trillion identical, calibrated

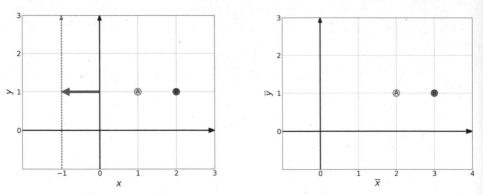

FIGURE 1.3. The separation of two points in space is independent of where we place the origin. Left: A and B are separated by a distance of 1, and we propose shifting (translating) the *y*-axis one unit to the left. Right: The separations between the two points (but not the distances to the origin) remain unchanged in the new (barred) frame.

timepieces spaced out in a predetermined configuration. Those trillions of physicists won't produce a great work of literature, but they *will* be able to tell you where and when a particle is.

As a practical matter, this frame of reference gives you an origin; *x*, *y*, and *z* axes; and (as we'll see later) a running clock. No matter how a particle moves about, within a particular frame of reference, we can unambiguously specify where that particle is.

There is nothing magical or absolute about our choice of reference frame, but once chosen, we'd better make sure that all of our calculations are done using the *same* origin and coordinate axes. There are tricks, naturally, to converting from one frame to another—from figuring out *relative* measurements from frame to frame (Relative! Get it? This isn't a digression; it's the main event)—but you need to know that you are doing those transformations.

1.3.2 TRANSLATION

As the simplest possible case, consider two different reference frames with slightly different origins (see figure 1.3). If you move the *y*-axis one unit to the left, we can make two different sets of measurements. Essentially, we're adding a simple offset to all measurements:

$$\bar{x} = x + 1; \quad \bar{y} = y,$$

where we use bars over variables to indicate two different sets of measurements.

Just to make things absolutely clear, we're imagining two entirely different sets of observers making measurements. One set of observers (the "unbarred" frame) measures the positions, velocities, forces, and other vectorial quantities. The other frame ("barred") measures the same particles but might get different numbers.

As a classic example in mechanics, we know that the potential energy of a mass near the Earth is $U = mgy$, where m is the mass of the particle in question, g is the gravitational constant near the surface of the Earth, and y is the height above the ground. But there's a quirk: the exact value of this energy depends on precisely where you've decided the ground is. As long as you're consistent about the zero point throughout your problem, you're in good shape, because energy calculations involve only the *change* in potential energy.

We can generalize how changing the origin will affect vectors. Returning to the example above, in the original reference frame, particle A is at

$$x_A^i = \begin{pmatrix} 1 \\ 1 \end{pmatrix},$$

while in the "barred" reference frame, it is at

$$x_A^{\bar{i}} = \begin{pmatrix} 2 \\ 1 \end{pmatrix}.$$

Instead of putting the bar over the x, we put the bar over the index as an indication that we're talking about directions in two different frames. (They're still numbered 1, 2, 3 in the new frame, however.)

Regardless of the frame, the distance between the two particles is 1. This is what is known as an **invariant**—a quantity that is exactly the same from one reference frame to another. The distance between two points is an invariant. The distance from the origin is not. Invariants are going to be one of the most important ideas in relativity.

Translations can be more generally written in terms of

$$x^{\bar{i}} = x^i + \Delta^i,$$

and the laws of physics are totally immune to them.

Consider Newton's (nonrelativistic) second law, in which we compute the components of force, F^i, on a particle:

$$F^i = m\frac{d^2 x^i}{dt^2}.$$

What happens if we try to compute the force $F^{\bar{i}}$ in another frame? Assuming it follows the same formula (it does), and assuming the frames are simply displacements from one another (they are), we get

$$F^{\bar{i}} = m\frac{d^2 x^{\bar{i}}}{dt^2}$$

$$= m\left(\frac{d^2 x^i}{dt^2} + \cancel{\frac{d^2 \Delta^i}{dt^2}}^{\,0} \right)$$

$$= F^i.$$

Or, to put it another way, forces are independent of where you put the origin.

There's a deeper interpretation to this, by the way. If you imagine Newton's laws to be the framework of physical laws, then we're basically arguing that physical laws are completely independent of position—you can't specify your position unambiguously, after all.

And if that's the case, then the next logical interpretation is that you can't do any experiment that tells you where you are in the universe. Or, to close out the line of reasoning, it means that absolute position is meaningless. The universe has no center, no edge, or anything else. Cosmologists have a term for this: the **cosmological principle**. It states, in essence, that wherever you go, there you are.

There's also a principle in mechanics that comes into play. In 1918, mathematician Emmy Noether [33] (figure 1.4) derived an important principle of symmetry. She found that symmetries in physical law give rise to conserved quantities. For example, if your physical law is independent of x, then linear momentum in the x-direction will necessarily be conserved. On the other side, if you do physics near the surface of the Earth, energy is explicitly a function of y, so the y component of linear momentum will *not* be conserved.

1.3.3 ROTATION

Translations aren't the only possible coordinate transforms. Which direction is x and which is y has no absolute meaning in physics. Start with the same two points we used above. We could imagine rotating the entire set of coordinate axes by a fixed amount. This, too, produces a new frame of reference (figure 1.5).

You may have done this sort of calculation before if you've ever had to compute the components along an inclined plane, but I'd like to take a closer

FIGURE 1.4. Amalie "Emmy" Noether (1882–1935), seen here around 1910, developed a core idea in mathematical physics that symmetries ultimately lead to conservation laws. Earth's attraction to the Sun, for example, is independent of direction, and while that doesn't manifest in a circular orbit, it does result in conservation of angular momentum.

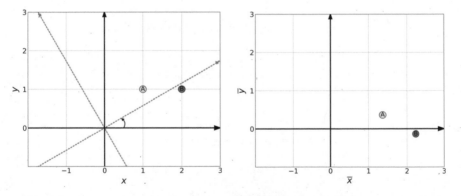

FIGURE 1.5. Left: Two points plotted in a two dimensional plane. A rotation of the coordinate axis is superimposed. Right: The coordinates in the rotated frame, $x^{\bar{i}}$, rotated by 30 degrees.

look. In terms of an equation, rotating the coordinates amounts to

$$\bar{x} = x \cos \theta + y \sin \theta \qquad (1.7)$$

$$\bar{y} = y \cos \theta - x \sin \theta.$$

In this particular worked example, I've rotated the frame by 30 degrees (the old standard for trigonmetric simplicity) and solved for the coordinates of point A as

$$x_A^i = (2, 1) \Longrightarrow x_A^{\bar{i}} = \left(\frac{2\sqrt{3}+1}{2}, \frac{\sqrt{3}-2}{2} \right),$$

with a similar tranformation for the other point.

While the transformed coordinates seem very different from the original, certain quantities have remained invariant. For instance, you can check using the Pythagorean theorem that point A is still $\sqrt{2}$ units from the origin, and B is still $\sqrt{5}$ from the origin. You can also check that the two are still one unit from each other.

There's a straightforward way of converting between one frame of reference and another. A **transformation matrix** can be defined as

$$\Lambda^{\bar{i}}_{\ i} \equiv \frac{\partial x^{\bar{i}}}{\partial x^i}. \tag{1.8}$$

In our case, our transformation is defined by equation 1.7 so, for instance,

$$\frac{\partial \bar{x}}{\partial x} = \cos \theta$$

and similarly for the other three derivatives. This yields the matrix

$$\Lambda^{\bar{i}}_{\ i} = \begin{pmatrix} \cos \theta & \sin \theta \\ -\sin \theta & \cos \theta \end{pmatrix}. \tag{1.9}$$

If we were writing this in terms of linear algebra, we'd say that $\bar{\bar{x}} = \Lambda \bar{x}$, but we can use the Einstein summation convention to write the same thing:

$$x^{\bar{i}} = \Lambda^{\bar{i}}_{\ i} x^i.$$

We can go from any frame to any other frame.

Right away, you'll notice some interesting things about the rotation matrix. For instance, if you plug in $\theta = 0$, you get $\begin{pmatrix} 1 & 0 \\ 0 & 1 \end{pmatrix}$, also known as the **Identity matrix**. That is, if you rotate by zero degrees, you get the matrix you start with.

The transformation allows us to compute the metric in a different rotated frame. Working out all of the gory details,

$$g_{\bar{i}\bar{j}} = \sum_i \sum_j \Lambda^i_{\bar{i}} \Lambda^j_{\bar{j}} g_{ij}$$

$$= \begin{pmatrix} \Lambda^1_{\bar{1}} \Lambda^1_{\bar{1}} g_{11} + \Lambda^2_{\bar{1}} \Lambda^2_{\bar{1}} g_{22} & \Lambda^1_{\bar{1}} \Lambda^1_{\bar{2}} g_{11} + \Lambda^2_{\bar{1}} \Lambda^2_{\bar{2}} g_{22} \\ \cdots & \cdots \end{pmatrix}$$

$$= \begin{pmatrix} \cos^2 \theta \times 1 + \sin^2 \theta \times 1 & \cos \theta(-\sin \theta) \times 1 + \sin \theta \cos \theta \times 1 \\ \cdots & \cdots \end{pmatrix}$$

$$= \begin{pmatrix} 1 & 0 \\ 0 & 1 \end{pmatrix}.$$

I told you that invariants were going to be important, and here we have one! The metric is invariant under both rotations and translations. This is a big deal. It is the reason you're allowed to simply rotate the frame and solve problems component by component in inclined plane problems.

1.4   A Very Brief Introduction to Group Theory*[†]

You may be wondering how I was able to just give you equation 1.9 and have it produce an unchanged metric. Part of this is that I've done this calculation before, but this is where the mathematics of **group theory** help us. A **group** is one of those ideas in mathematics that is exactly what it sounds like: it's a collection of related objects. Those objects can be anything, but in mathematical physics, they are almost always operations.

As a simple example, consider an equilateral triangle with the vertices labeled (for bookkeeping purposes only) as A, B, and C, as shown in figure 1.6. Some manipulations of a triangle leave it looking the same as it did before; in the case of equilateral triangles, reflections and rotations by 120 degrees have this property. This set of manipulations is a group.

A group of operations has a number of properties, and rotations and translations satisfy all of them.

1. *Closure.* Multiply two different elements of a group and you get a third element. In our case, rotating twice counterclockwise is the same as rotating once clockwise. Three rotations is the same as doing nothing.
2. *Associativity.* You can group the group elements however you like. (Ignore this for now.)
3. *Identity.* Some group element does nothing. As I noted, this is the same as rotating three times.

[†]Sections denoted with an asterisk may be omitted on a first reading or in a condensed course without significant disadvantage.

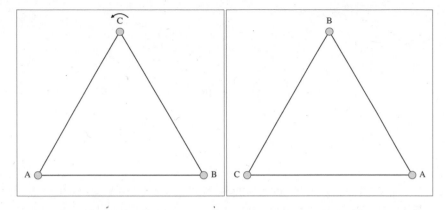

FIGURE 1.6. A simple equilateral triangle can be manipulated in a number of ways, including rotation, reflection, or doing nothing at all, and it will look the same as it did before. The labeled vertices are for reference, but as a practical matter, before and after the rotation, the triangle is precisely the same.

4. *Inverse.* You can find an element that exactly undoes the transformation. For the triangle, rotations clockwise and counterclockwise are inverses of one another.

Just as we can make a group that leaves an equilateral triangle the same, we can make an infinite group for all of the possible rotations that leave a two-dimensional metric unchanged:

$$R(\theta) = \begin{pmatrix} \cos\theta & \sin\theta \\ -\sin\theta & \cos\theta \end{pmatrix},$$

for every possible real value of θ. And we can verify that it satisfies all of the group rules: For instance, closure: Rotate by θ_2 and then by θ_1 on a vector, and this is mathematically the same as multiplying by

$$\begin{aligned}
R(\theta_1)R(\theta_2) &= \begin{pmatrix} \cos\theta_1 & \sin\theta_1 \\ -\sin\theta_1 & \cos\theta_1 \end{pmatrix} \begin{pmatrix} \cos\theta_2 & \sin\theta_2 \\ -\sin\theta_2 & \cos\theta_2 \end{pmatrix} \\
&= \begin{pmatrix} \cos\theta_1\cos\theta_2 - \sin\theta_1\sin\theta_2 & \cos\theta_1\sin\theta_2 + \sin\theta_1\cos\theta_2 \\ -\sin\theta_1\cos\theta_2 - \cos\theta_1\sin\theta_2 & -\sin\theta_1\sin\theta_2 + \cos\theta_1\cos\theta_2 \end{pmatrix} \\
&= \begin{pmatrix} \cos(\theta_1 + \theta_2) & \sin(\theta_1 + \theta_2) \\ -\sin(\theta_1 + \theta_2) & \cos(\theta_1 + \theta_2) \end{pmatrix} \\
&= R(\theta_1 + \theta_2).
\end{aligned}$$

If you rotate twice, we've just found it's the same as rotating once by the sum of the two angles. Or, in other words, two rotations (each a member of the rotation group) produce a third member. *Closure!*

The other rules are simpler. For instance, plug in $\theta = 0$ to the matrix definition, and you get the identity matrix. Do nothing, or *Identity*.

The group of rotations in two dimensions (the ones we've been using) is known as SO(2):

- S: "Special." What specialness means in this case is that you can take the **determinant** of the transformation matrix, and it's always 1. Indeed, for rotations, this is necessarily the case; $\cos^2\theta + \sin^2\theta = 1$ regardless of the angle you choose.
- O: "Orthogonal," because the transposition (flipping the matrix diagonally) undoes the rotation. We can even prove this. Recall that cosines are odd functions, and sines are even. That is,

$$\cos(-\theta) = \cos(\theta); \quad \sin(-\theta) = -\sin(\theta).$$

  Reversing the direction—inverting the rotation—just flops the position of the minus sign. Or, to put it another way, going *back* from the barred frame to the unbarred:

$$\Lambda^i_{\bar{i}} = \begin{pmatrix} \cos\theta & -\sin\theta \\ \sin\theta & \cos\theta \end{pmatrix}. \tag{1.10}$$

  We've found our *inverse!*
- 2: Because it's two-dimensional. In case you hadn't figured out the pattern, the set of matrices that would let you rotate in three dimensions would be SO(3).

Knowing that our metric (for instance) is invariant under a group of rotations allows us to gloss over a whole lot. Having proven all of this, we can simply say that any equations true in one frame will be equally valid in another rotated frame. But as we're about to see, this isn't true with all transformations.

1.5   Not All Transformations Are Metric Invariant*

We chose to show that translations (moving the origin) and rotations left the metric unchanged. This was by design. We're deeply interested in transformations that don't change the metric, because those are also going to be the ones that keep our physical laws intact.

I briefly introduced, but didn't prove, Noether's theorem for the case of translational invariance: if the laws of physics (and for our purposes, this means the metric) don't change under translations, then you've got a conservation of linear momentum. By the same token, rotations leave the metric the same, which means that the laws of physics are the same in all directions. The only reason we identify "down" or "north" on the Earth is because of the relative position of the center of the Earth to our feet or of the North Pole.

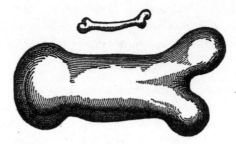

FIGURE 1.7. From Galileo's *Dialogue Concerning Two New Sciences*, in which he speculates on the existence of giants. As noted above, because of the increased stress on giant bones, they would necessarily be proportionately much wider than ordinary human bones.

If the whole solar system were turned 90 degrees with respect to the center of the Galaxy (as it will be in about 50 million years [18]), we'd be none the wiser.

As with translations, the rotational invariance of physical law also predicts a conservation law from Noether's theorem: conservation of angular momentum. And since the laws themselves are rotationally invariant, angular momentum is a globally conserved quantity.

Those two choices, translation and rotation, are special. But not all transformations work the same way.

1.5.1 SCALING

Galileo was among the first physicists to think deeply about the idea of symmetry in physical laws. He recognized symmetries of translation and rotation but was more cautious about scale [19]:

> An oak two hundred cubits high would not be able to sustain its own branches
> if they were distributed as in a tree of ordinary size; and that nature cannot
> produce a horse as large as twenty ordinary horses or a giant ten times taller
> than an ordinary man.

He even helpfully included a diagram of what giant bones might look like (figure 1.7).

You might even have a sense as to why giants are scarce. Gravity follows an inverse square law, which means that if you halve all distances, then the force of gravity should increase by a factor of four.

But we can use our newfound sense of invariants to figure out why scale matters. As good scientists, we do most of our calculations in meter-kilogram-second (MKS) units. But suppose we wanted to do calculations in terms of centimeter-gram-second (CGS) units—turning all distances in meters to distances in centimeters. In terms of calculations, this means that our new frame would give positions and velocities 100 times larger than if we had calculated in meters.

In terms of a transformation to a barred (centimeter) frame,

$$\bar{x} = 100x; \quad \bar{y} = 100y,$$

or, in terms of the transformation matrix,

$$\Lambda^{\bar{i}}_{\ i} = \begin{pmatrix} 100 & 0 \\ 0 & 100 \end{pmatrix}.$$

We may then invert:

$$\Lambda^{i}_{\ \bar{i}} = \begin{pmatrix} \frac{1}{100} & 0 \\ 0 & \frac{1}{100} \end{pmatrix}.$$

Already, this appears very different from the rotation matrix. You can see by inspection that the barred frame will always have larger coordinates than the unbarred one. That was the whole point. But we can also follow the example from rotation to find the metric in the barred frame:

$$g_{\bar{i}\bar{j}} = \begin{pmatrix} \frac{1}{10,000} & 0 \\ 0 & \frac{1}{10,000} \end{pmatrix}.$$

New frame, and we have a new metric! This, as Galileo noted, means that scale invariance is not a symmetry of nature.

But there's also an important corollary to all of this. For scalars—quantities where all of the upstairs and downstairs indices cancel one another—the final product is completely unchanged. For instance, a point at

$$x^{i} = \begin{pmatrix} 3 \\ 4 \end{pmatrix}$$

has distance squared of $dl^2 = 25$ from the origin.

If we convert to our centimeter frame, then the new coordinates are

$$x^{\bar{i}} = \begin{pmatrix} 300 \\ 400 \end{pmatrix}.$$

I'll leave it as a one-step exercise to show that $dl^2 = 25$ in the new frame as well.

A scalar is a scalar and produces the same value, no matter how wacky a frame you choose.

1.5.2 POLAR COORDINATES

We'll do one more coordinate transformation, one that you've likely seen (or will see) in the context of orbital mechanics: polar coordinates. In this case, we have two coordinates:

$$x^{\bar{i}} = \begin{pmatrix} r \\ \theta \end{pmatrix},$$

where

$$x = r\cos\theta$$
$$y = r\sin\theta,$$

and so the transformation matrix is

$$\Lambda^i{}_{\bar{i}} = \begin{pmatrix} \cos\theta & -r\sin\theta \\ \sin\theta & r\cos\theta \end{pmatrix}. \tag{1.11}$$

We can also transform the metric to polar coordinates. Rather than get totally bogged down, let's just compute the diagonal elements of the polar coordinate metric:

$$g_{\bar{1}\bar{1}} = \Lambda^1{}_{\bar{1}}\Lambda^1{}_{\bar{1}}\cancelto{1}{g_{11}} + \Lambda^1{}_{\bar{1}}\Lambda^2{}_{\bar{1}}\cancelto{0}{g_{12}} + \Lambda^2{}_{\bar{1}}\Lambda^1{}_{\bar{1}}\cancelto{0}{g_{21}} + \Lambda^2{}_{\bar{1}}\Lambda^2{}_{\bar{1}}\cancelto{1}{g_{22}}$$
$$= (\cos\theta)^2 + 0 + 0 + (\sin\theta)^2$$
$$= 1.$$

A lot of work to end up with a very simple result. However, $g_{\bar{2}\bar{2}}$ is a little more interesting:

$$g_{\bar{2}\bar{2}} = \Lambda^1{}_{\bar{2}}\Lambda^1{}_{\bar{2}}\cancelto{1}{g_{11}} + \Lambda^1{}_{\bar{1}}\Lambda^2{}_{\bar{2}}\cancelto{0}{g_{12}} + \Lambda^2{}_{\bar{2}}\Lambda^1{}_{\bar{2}}\cancelto{0}{g_{21}} + \Lambda^2{}_{\bar{2}}\Lambda^2{}_{\bar{2}}\cancelto{1}{g_{22}}$$
$$= (-r\sin\theta)^2 + 0 + 0 + (r\cos\theta)^2$$
$$= r^2.$$

Expanding all of the elements gives the full polar metric, we get

$$g_{\bar{i}\bar{j}} = \begin{pmatrix} 1 & 0 \\ 0 & r^2 \end{pmatrix}. \tag{1.12}$$

We're not going to use polar coordinates in this book (which is why this section has an asterisk). However, if you want to get a sense of how this result is used in ordinary dynamics, consider a particle in orbital motion. The "velocity" components can be written as

$$V^{\bar{i}} = \begin{pmatrix} \dot{r} \\ \dot{\theta} \end{pmatrix},$$

where we use the convention $\dot{r} = \frac{dr}{dt}$. Thus, to calculate the total velocity, we calculate

$$|\vec{V}|^2 = g_{\bar{i}\bar{j}} V^{\bar{i}} V^{\bar{j}} = \dot{r}^2 + r^2 \dot{\theta}^2,$$

which is nothing more than the quadratic sum of the radial and tangential velocity! And the tangential velocity just falls out as $r\dot{\theta}$.

Looking Forward

We've done a lot of work writing down spatial coordinates and doing rotations and translations and such. And for what?! To produce results that you already knew? To write down the Pythagorean theorem in a slightly more elegant fashion? Our goals are loftier than that. We introduced all of this notational work *because* you already have an intuition of how it's supposed to work. You already know what the answers are supposed to be, so you're less bogged down by the mechanics of messing with coordinates in space.

In the next chapter, we'll add a time dimension to our analysis, and there, too, you'll have an intuition—but it will turn out that your intuition will be wrong. You are a product of the low-speed, nonrelativistic world. You'll have to learn how space and time interact all over again, but fortunately, the mathematical tools we developed in this chapter will provide a road guide for doing so.

1.6 Problems

1. For each of the following "tensor equations," use the tensor rules (available in one place in the appendix) to determine whether either the equation is valid (whether or not the physics is correct), or, if it is not valid, why not?
   (a) $x^i = y^j$
   (b) $ma^i = F^i$
   (c) $q^i = c_i$
   (d) $A^{ij} = 2B^{ij} + C^{ij}$

2. In the rotation example in the text (eq. 1.7), we calculate the positions of two points in both an original frame, $x^i_{A,B}$, and a rotated frame, $x^{\bar{j}}_{A,B}$. For each frame of reference, compute the following:

(a) The distance of point A from the origin in the unbarred and barred frames

(b) The coordinates of point B in the barred frame

(c) The distance of point B from the origin in both the unbarred and barred frames

(d) The distance of point A from point B in both the unbarred and barred frames

3. Consider the following two vectors

$$A^i = \begin{pmatrix} -4 \\ 3 \\ 0 \end{pmatrix}; \quad B^i = \begin{pmatrix} 1 \\ -2 \\ 0 \end{pmatrix}$$

with a Euclidean metric:

$$g_{ij} = \begin{pmatrix} 1 & 0 & 0 \\ 0 & 1 & 0 \\ 0 & 0 & 1 \end{pmatrix}.$$

(a) Sketch both vectors (in the xy-plane only).

(b) Calculate $g_{ij}A^iA^j$. Note that this can also be expressed as $\vec{A} \cdot \vec{A} = |\vec{A}|^2$. Based on this calculation, how long is vector A^i?

(c) Calculate $g_{ij}A^iB^j$.

(d) Consider rotation of the coordinate axes $\theta = \pi/6$ (or, equivalently, 30 degrees) given by the transformation

$$\Lambda^{\bar{i}}_{\ i} = \begin{pmatrix} \cos\theta & \sin\theta & 0 \\ -\sin\theta & \cos\theta & 0 \\ 0 & 0 & 1 \end{pmatrix}.$$

What are the values of $A^{\bar{i}}$ in this new frame?

(e) What are the values of $B^{\bar{i}}$ in this new frame?

(f) Sketch both vectors with the new, rotated axes.

(g) Remembering that the metric doesn't change with rotations, calculate $g_{\bar{i}\bar{j}}A^{\bar{i}}B^{\bar{j}}$.

4. Consider a rank-2 tensor Q and a vector u,

$$Q^{ij} = \begin{pmatrix} 4 & 2 & 1 \\ -5 & 0 & 6 \\ 2 & 0 & 0 \end{pmatrix}; \quad u^k = \begin{pmatrix} 2 \\ 0 \\ -2 \end{pmatrix},$$

where in the tensor, the second index, j, represents the column numbers.

(a) What are the elements of the 1-form $g_{ij}u^j$?

(b) What is the value of u^iu_i?

(c) What are the components of $Q^{ij}u_j$?

(d) What is the value of $Q_{ij}Q^{ij}$?

5. Consider a coordinate transformation where $\bar{x} = 2y$ and $\bar{y} = x$.
   (a) What is the transformation matrix $\Lambda^{\bar{i}}_{\ i}$?
   (b) What is the inverse transformation matrix $\Lambda^{i}_{\ \bar{i}}$?
   (c) What is the metric in the barred frame?
6. As noted in the text, the general group for rotating in three dimensions is called SO(3). The rotation around the z-axis is shown in problem 3d, above.

   Following that example, how would you construct a rotation around the y-axis? What is the determinant of that matrix?

   (Note: "S" stands for special, which should give you the answer right away.)
7. In the derivation of the polar coordinate metric, we skipped over the calculation of the off-diagonal terms. Following the example for the diagonal terms, compute $g_{\overline{12}}$ explicitly.

FURTHER READINGS

While technical references are generally cited at the end of the book, at the end of each chapter, I'll list some additional nontechnical or semitechnical works that you might want to read to supplement your historical or philosophical background.

- Leonard Mlodinow, *Euclid's Window: The Story of Geometry from Parallel Lines to Hyperspace* (Free Press, 2002). While an interested student might want to (following Einstein) delve into Euclid directly, Mlodinow provides a very nice treatment of how geometry evolved and influenced our understanding of the physical world.
- Dwight E. Neuenschwander, *Emmy Noether's Wonderful Theorem.* (Johns Hopkins University Press, 2011). While I only briefly discuss the impact of Noether in this chapter, her work is a cornerstone of modern physics. I strongly recommend this work, which provides a nice mix of the mathematical and the historical.
- Edvin F. Taylor and John Archibald Wheeler, *Spacetime Physics* (W. H. Freeman, 1992). This is a great classic text to introductory special relativity, taking an almost orthogonal approach to the present volume. It is highly conceptual, with a focus on trying to build up a physical intuition.
- J. L. Synge and A. Schild, *Tensor Calculus* (Dover, 1949). While much of the book focuses on curved spaces and differential geometry, the interested reader will find the first two chapters, which provide a general introduction to tensors, especially useful.

Spacetime

The Michelson-Morley apparatus, a stone slab mounted on liquid mercury. Albert Michelson and Edward Morley showed in 1887 [28] that the speed of light was independent of the relative state of motion of source and detector. Courtesy Case Western Reserve University.

We started our study of relativity with a discussion of space, coordinates, and transformations. We laid out points and particles in the world, and there they stayed. But this is a physics text; things are going to move. And that means we're going to move from space to **spacetime**. Spacetime—and relativity more generally—is predicated on the idea that time is just another coordinate, along with length, height, and depth.

At first blush time *seems* very different than the other coordinates. You can jump to the left, and you can jump to the right, but as the poet (and mathematician!) Omar Khayyám [14] put it:

> The Moving Finger writes; and, having writ,
> Moves on: nor all thy Piety nor Wit
> Shall lure it back to cancel half a Line,
> Nor all thy Tears wash out a Word of it.

In addition to its one-way nature, time presents a confounding distinction from space: left-right and up-down measurements can be made in meters, but time is measured in seconds or years. We need some way to convert between space and time if we're somehow going to be able to use them interchangeably.

It is surprisingly difficult to measure lengths in a precise way. For over 70 years, up until 1960, a platinum-iridium rod was kept in a vault in France for absolute reference, [32] but its 31 copies were still too few to effectively measure everything that needed measuring.

It's easier to accurately and unambiguously measure time. The approach accepted by the U.S. National Institute of Standards and Technology (NIST) and by most countries [1] is to multiply the period of atomic oscillations by a large factor. In particular, time measurement is based on the frequency of the so-called *hyperfine* transition frequency of ground state cesium. The details aren't important, but, fundamentally, there's an oscillation, and when a cesium atom oscillates approximately 9.192 billion times, that's a second.

2.1 The Speed of Light

Having defined units of time, we can relate them to units of space via the **speed of light**, c. In the early 1600s, Galileo tried to measure the speed of light. He and an assistant would stand on distant hills holding lamps with shutters in them. Galileo would open and close his lamp quickly, and his assistant was to do the same as soon as he saw the light from Galileo's lamp. The amount of time that passed between when Galileo opened his lamp and when he saw his assistant's lamp would, in principle, give him twice the light travel time between the two hills. Galileo conceded that his results were, at best, inconclusive: [19]

> In fact I have tried the experiment only at a short distance, less than a mile, from which I have not been able to ascertain with certainty whether the appearance of the opposite light was instantaneous or not; but if not instantaneous it is extraordinarily rapid.*

In 1676, Ole Rømer, a Danish astronomer working in Paris, was the first to make an accurate astronomical measurement of the speed of light. [38] While he was trying to measure the period of Io (the nearest big moon to Jupiter), he found something interesting: Although the orbit of Io should behave like a clock, when Earth was closest to Jupiter, Io seemed to run a bit ahead of schedule. Six months later, when Earth was furthest from Jupiter, Io seemed to be behind schedule. Rømer surmised that the difference came about from the finite speed of light having to travel further. By Rømer's calculation, it

---

*Indeed, it is claimed (perhaps apocryphally) that the abbreviation c stems from the Latin *celeritas*, for swiftness.

should take about 22 minutes for light to traverse the diameter of Earth's orbit (a more accurate modern estimate would be about 17 minutes).

A few years earlier, in 1672, Giovanni Domenico Cassini and Jean Richer, and, independently, John Flamsteed, were able to measure the distance to Mars, [30] setting the scale of the Solar System—and thus producing an absolute measure of the **astronomical unit**, the distance from Earth to the Sun. Their results weren't perfect; they were off by about 10%, and combining the distance measurements with Rømer's timing gave a speed of light accurate to within about 30% of the modern value.

Over time, astronomical measurements of the speed of light got better and better (ultimately converging to within 1% of the modern value), but the first accurate terrestrial measurements—the ones that ultimately pegged our modern speed of light—had to wait until the mid-nineteenth century. Hippolyte Fizeau [16] and Léon Foucault [17] designed experiments to measure an authoritative and absolute estimate of the speed of light. Conceptually, their devices were similar to Galileo's approach, but used timed, rotating cogs and mirrors rather than the fallibility of two humans. Fizeau and Foucault found c within a few percent of the modern value:

$$c \equiv 299,792,458 \, \text{m/s}. \tag{2.1}$$

This relation is *exact*. We noted earlier that it's easier to measure intervals of time than length. Thus, a second is defined in terms of atomic transition frequencies, but, following NIST and other international standards, a meter is simply *defined* as $1/299,792,458$ of the distance that light travels in a second. As a practical matter, the approximation $c \approx 3 \times 10^8 \, \text{m/s}$ will work just fine.

We haven't proven it yet, but the thing that we call the speed of *light* is really the speed of *all* massless particles, of which light is simply the most famous. There's a corollary to this: all massive particles travel slower than c.

2.1.1 NATURAL UNITS

We've spent so long talking about the speed of light for two reasons: 1) We will see it is the speed limit of the universe, and 2) it's really useful for converting between time and space. We don't even need to worry about the conversion factors between time and space most of the time. We can simply introduce a unit of distance, the **light-second**, which is simply

$$1 \, \text{ls} = c \times (1 \, \text{s}) \approx 3 \times 10^8 \, \text{m},$$

or roughly the distance to the moon. A light-year is

$$1 \, \text{ly} = c \times (1 \, \text{yr}) \approx 9.46 \times 10^{15} \, \text{m}.$$

TABLE 2.1   Some useful natural unit conversions.

Time	Length	Mass	Energy
1 s	$1 \, ls = 3 \times 10^8 \, m$	1 kg	$9 \times 10^{16} \, J$
1 yr	$1 \, ly = 3.15 \times 10^{15} \, m$	$m_e = 9.11 \times 10^{-31} \, kg$	511 keV
3.3 ns	1 m	$m_P = 1.67 \times 10^{-27} \, kg$	938 MeV
		$1.11 \times 10^{-17} \, kg$	1 J

Let's just rip the Band-Aid off and get rid of the "light" in light-year, asserting that space and time have the same units:*

$$[L] = [T].$$

Setting distance and time to the same units means that velocities are dimensionless—expressed as a fraction of the speed of light—and, by construction,

$$c = 1 \tag{2.2}$$

in these units. These are called **natural units**, and they'll serve as a very useful tool in keeping our relativistic equations clean. Setting $c = 1$ neatens up *every* equation involving velocities. Even if we didn't already know the most famous equation in physics, dimensionally, energy is mass times velocity squared. Thus, in natural units:

$$[E] = [M].$$

If you want to turn a mass into an energy, simply multiply it by c^2. I've put some useful unit conversions in table 2.1.

2.1.2   NATURAL AND GEOMETRIZED UNITS*

I made the assertion that $c = 1$ is known as "natural units," but that isn't quite right. If you're familiar with the MKS (meter-kilogram-second) system, you'll note that there are three "fundamental" units from which composite units can be defined. A joule, for instance, is a kg m/s². By setting $c = 1$, we allow a length-time equivalence and a mass-energy equivalence, but there's still another degree of freedom needed to allow those two to be equal to *each other*.

There are two main options for setting that degree of freedom: **natural** and **geometrized** units.

---

*If you've never seen this notation before, the hard brackets in this notation mean "the units of." Thus $[L]$ simply means the units of length, which would be meters in a standard meter-kilogram-second (MKS) calculation.

Natural Units

Quantum mechanics, as you likely know, dominates on the scale of the very small. For instance, for a particle of mass m, we can define a **Compton scale** as

$$\lambda_C = \frac{\hbar}{mc},$$

where $\hbar$ is the reduced Planck constant, a sign that quantum mechanics is at play. Quantum mechanics predicts, among much else, that particle-antiparticle pairs pop into and out of existence, and the Compton scale gives us a sense as to how long those pairs last. The larger the mass, the smaller the length or time scale.

The Compton relationship simplifies considerably if we include a second convention for natural units, setting the reduced Planck constant, $\hbar$, as

$$\hbar = 1.$$

This means

$$[E] = [T]^{-1}.$$

Nuclear physics is smaller scale and higher energy than atomic or molecular physics. To get a sense of the scales involved, consider the following energy-time-length conversions in natural units:

| | | Natural Unit | |
Scale	Energy	Time	Length
Planck	1.2×10^{19} GeV	5.4×10^{-44} s	1.6×10^{-35} m
Baryons	1 GeV	6.5×10^{-25} s	1.9×10^{-16} m
Molecular	10 eV	$6,5 \times 10^{-17}$ s	1.8×10^{-8} m

Perhaps unsurprisingly, using only dimensional analysis, the length scale associated with baryons (protons and neutrons) corresponds roughly to a femtometer, the canonical size of an atomic nucleus. Likewise, using energies in the 10s of eV range, we find length scales on order of a nanometer—roughly the size of an electron cloud.

Geometrized Units

At the other extreme, we have the realm of general relativity. The archetypal general relativistic object is a black hole with a radius of

$$R_S = \frac{2GM}{c^2}.$$

For the uninitiated, this is the **Schwarzschild radius**, the boundary beyond which nothing can escape, not even light.

Our definition of the Schwarzschild radius (and indeed, all of general relativity) simplifies considerably if

$$G = 1,$$

the second criteria (after $c = 1$) for geometrized units. In these units,

$$[E] = [T].$$

Since distance, time, energy, and mass are all the same unit, unit analysis in geometrized units is especially simple. As with natural units, let's take a look at some relevant scales in geometrized units:

		Geometrized Unit	
Scale	Energy	Time	Length
Sun	$M_\odot = 2 \times 10^{30}$ kg	5×10^{-6} s	1.5 km
Earth	$M_\oplus = 6 \times 10^{24}$ kg	1.5×10^{-11} s	.45 cm
Planck	1.2×10^{19} GeV	5.4×10^{-44} s	1.6×10^{-35} m
Baryons	1 GeV	4.5×10^{-63} s	1.3×10^{-54} m

You might notice that, for the Planck scale, the geometrized and natural units are exactly the same. That's the point. For anything shy of stars, the geometrized scales are so tiny as to be essentially ignorable. That, too, is the point. It's a testament to the fundamental weakness of gravity.

2.1.3 THE INVARIANT SPEED OF LIGHT

We've spent a fair amount of time talking about the speed of light, but what we haven't yet established is that it is *the* speed of light—that there's just one and no other.* Is light just a matter of perspective? In 1887, Albert Michelson and Edward Morley performed their famous interferometry experiment. [28]†

Their starting point was simple—and completely understandable. Light was known to exhibit wave properties of diffraction and interference. And *other* wave phenomena like sound waves, water waves, and seismic waves all required media through which to travel, as was proposed by Christiaan Huygens two centuries prior [22].

---

* At least, that is the case in a vacuum.

† Technically, there were several generations of experiment, but the definitive one was conducted in 1887.

Huygens suggested that light, like other waves, traveled through a medium, the **luminiferous æther**. Even if we couldn't see it directly, the æther should produce some detectable effects. For instance, if a light source is emanating in the same direction as the motion of the æther, we'd expect light to travel faster than average, while sources propagating through a "tailwind" should clock a slower speed of light.

Imagine that the æther is moving at speed v in the direction of the propagation of the wave. Knowing nothing else, you'd expect that light should propagate at a speed of something like

$$c_{medium} = c \pm v,$$

where you get the + if the detector is traveling into the wind and − if it's moving away.

We can't know for certain whether we're moving with respect to the æther, but if you measured the speed of light six months apart, Earth's velocity will shift by 60km/s (going one way around the Sun rather than the other), so the Sun's rays would æither travel with the æther or against it.

Michelson and Morley found that it doesn't matter when or in what orientation you configure the device, $c = c$. This would turn out to be one of the foundational ideas in relativity.

2.2   Spacetime Diagrams

I noted at the top of this chapter that it is, in some ways, easier to measure time consistently and precisely than it is to measure distances. But with a constant speed of light, we can simply define the latter in terms of the former. As Hermann Minkowski noted in 1908 [29]:

> Henceforth space by itself, and time by itself, are doomed to fade away into mere shadows, and only a kind of union of the two will preserve an independent reality.

Welcome to the world of spacetime.

2.2.1   EVENTS, WORLDLINES, AND ALL THAT

Minkowski, who did more than just about anyone to conceptualize the equivalence of space and time, introduced the idea of spacetime diagrams as a way of visualizing the relationship between the two.

The universe, so far as we can tell, has three dimensions of space and one dimension of time. As a shorthand, we'll say that it has a 3 + 1

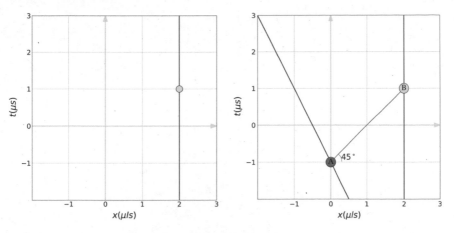

FIGURE 2.1. Successive construction of a spacetime diagram. Left: A single event in spacetime, at $t = 1\mu s$ and $x = 2\mu ls$, with the captain's world line running through it. In this particular case, the captain is stationary, which means that they're moving only in time and not in space. Right: Ship A shoots a light pulse (shown as a dashed line) toward the captain's ship (B). We know that it's a light beam because it makes a 45-degree angle with the horizontal. Meanwhile, ship A flies to the left at $0.5c$. We can see this directly as it travels $-2\mu ls$ in $4\mu s$.

dimensional spacetime. That's four dimensions overall, which is obviously too many to draw on a sheet of paper or to even visualize with virtual reality glasses.

Fortunately, for the overwhelming majority of our examples, we'll be just fine working in $1 + 1$ dimensional spacetime, and we *can* draw two dimensions on a sheet of paper. Spacetime diagrams have two axes: time (vertical) and space (horizontal). They are generally plotted with an equal aspect ratio and with space and time given the same units.

Each point in the diagram has a unique place and time and is known as an **event**. Rather than lay out abstract rules, we'll construct a spacetime diagram by example. Start by visualizing a spaceship captain looking at their watch (figure 2.1, left panel). This is an event that takes place at a specific point in space and a specific moment in time, and thus is just a dot in a spacetime diagram.

Pretty boring, no? While an event is a single point in the lifetime of the ship (and its captain), the ship itself existed in the past and continues existing into the future. The ship has a **worldline**, which extends into the past (down) and future (up).

In this case, the worldline is vertical. That's another way of saying that the velocity of the ship is zero. It's moving through time but not through space. As a practical matter, all nonrelativistic objects will appear to be more or less

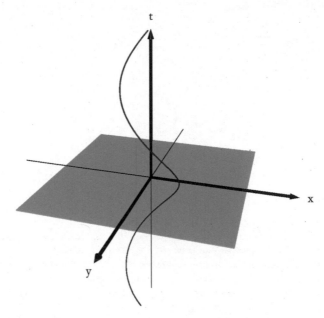

FIGURE 2.2. The trajectory of a particle in a 2 + 1 dimensional spacetime diagram. Time still points upward, but now we can see the motion of the particle in two spatial dimensions. We won't make a habit of working in more than one spatial dimension, but I figured you should get a sense of what it looks like. Alas, due to the restrictions of a flat page, we won't be drawing 3+1 dimensional spacetime diagrams.

vertical in a spacetime diagram because even very fast speeds (by human standards) are still much, much slower than the speed of light.

Finally, we can illustrate more complicated dynamics. Suppose we have a second ship (labeled A in figure 2.1), which shoots a laser at ship B. Lasers travel at the speed of light, which means that they traverse 1 light-second in space for every second of time. In spacetime diagrams, they are always drawn at 45-degree angles.

Spaceship A is simultaneously moving through space, and it's really truckin'. From $t = 0\,\mu$s to $t = 3\,\mu$s, it moves to the left by $\Delta x = -1.5\,\mu$ls, which means that it's moving at a velocity of

$$v = \frac{\Delta x}{\Delta t} = -0.5,$$

or half the speed of light to the left.

You'll draw and interpret a number of spacetime diagrams in the problems at the end of the chapter, but before we get to that, we should at least look at what a 2 + 1 dimensional spacetime diagram might look like. In figure 2.2, I draw a particle moving in a circular orbit around the origin in the xy-plane. It's moving fast, about 90% the speed of light, but at a constant speed.

2.2.2 THE FUTURE AND THE PAST

We haven't yet figured out the logistics, but the central idea in relativity is that space and time are, well, *relative*. But what does that mean? We'll see that two events in spacetime might be seen to take place in one chronological order according to one frame of reference and in the opposite order in another frame.

But on the other hand, we can make some definitive statements even now with the simple assertion that, if one event causes another, then it must unambiguously come first. The speed of light is the ultimate speed limit in the universe, so if you want to communicate with a friend or warn them of imminent danger, you only have a finite amount of time to do so. This is the nature of **causality**.

We've already noted that in a spacetime diagram light travels at 45-degree angles with respect to the horizontal, so at any particular point in spacetime (we'll use the origin), there is a **light cone** designating all of the events that could be exactly hit by a beam of light originating at the origin. Likewise, there's a past light cone indicating all possible events that could have influenced the origin. Anything outside of those cones can neither affect you at the origin (where both space and time are zero) nor be affected by you. Figure 2.3 features sketches of both a 1 + 1 spacetime diagram and a 2 + 1 version (which makes the "cone" part of light cone clearer).

2.2.3 WHY TIME IS DIFFERENT*

Unit analysis and the speed of light dictate that time is *like* space in many regards, but there's an important way in which it's different: Why can we move back and forth in space but not in time? That is, why is there an **arrow of time**?

There is a sense in which the future and the past don't really matter all that much, and for our work in this book, the direction of the arrow of time is more a curiosity than anything else. Indeed, the fundamental laws of physics are largely immune to the arrow of time.* As a simple example, consider a planetary orbit or a (classical) electron orbit around an atom. You could record either one, run the movie in reverse, and it would look as physically plausible as the forward version.

And yet, on the macroscopic scale, there is one law that seems to be violated if we switch past and future: the second law of thermodynamics. French

---

*The weak nuclear force is an important exception.

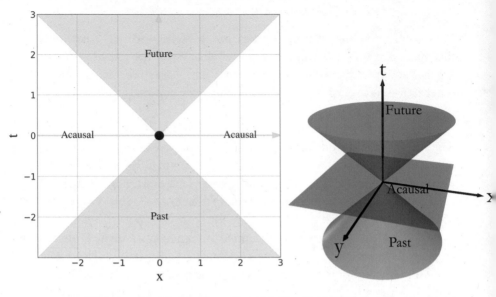

FIGURE 2.3. A view of the future and the past in $1 + 1$ (left) and $2 + 1$ (right) dimensions. An observer stands at the origin, and at $t = 0$, a light beam could reach them from the bottom cone, while emitted light would follow the upper cone. The interior of the lower cone could, in principle, affect the event at the origin, while events within the upper cone exist in the origin's future. Outside the cones, there's no causal connection.

engineer Nicolas Léonard Sadi Carnot is generally credited with devising the three laws of thermodynamics in 1824 [4]. While Carnot was focused on the transition of thermal energy to mechanical energy, in the modern interpretation, the second law involves the increase in entropy. Roughly speaking (and we may be quite rough, as the topic is outside the realm of this book), systems tend to become more disordered over time. Pool balls arranged in a neat triangle are in a high degree of order, and an initial break tends to scatter them. You'd be very surprised to find a random shot that resulted in a scatter of balls neatly arraying themselves back into a triangle.

2.3 The Minkowski Metric

2.3.1 4-VECTORS

We've seen a graphical approach to describing spacetime, but we also need to adjust our algebraic approach. In Euclidean space we had a convention where, for instance, x^i represented the components of a position vector. The index—any Roman index—i can take on the values of 1, 2, and 3, corresponding to x, y, and z.

We could add our time component wherever we like (indeed, Einstein put time as the fourth component), but the modern convention is to label the 0th component of a **4-vector** as time. To indicate that we're talking about spacetime, we use Greek indices. For instance, we might define the spacetime coordinates of a particle as

$$x^{\mu} = \begin{pmatrix} t \\ x \\ y \\ z \end{pmatrix}. \tag{2.3}$$

Every now and again, we're going to refer the 4-vector itself, rather than the components. Just as $\vec{v}$ will represent the entirety of a 3-vector, we'll use the boldfaced **x** to represent a 4-vector. The various notational forms of 4-vectors, 3-vectors, components, 1-forms, and so forth are collected in Appendix A.1.

As with 3-vectors, we're not restricted to position. For instance, consider the 4-vector

$$U^{\alpha} = \begin{pmatrix} 5/3 \\ 0 \\ 4/3 \\ 0 \end{pmatrix}.$$

We don't know much about what this 4-vector actually *means*. For the moment, it's just a list of four numbers rather than three. And this extends to other tensorial objects. If we encounter a rank-2 tensor, something labeled as $T^{\mu\nu}$, for instance, we'd know that it's a list of $4 \times 4 = 16$ numbers. We don't know what they mean or what they do. We just know that there are 16 of them.

2.3.2 INTERVALS

In the previous chapter, we found a couple of results from ordinary Euclidean geometry, the most important of which was the calculation of distance between two nearby points:

$$dl^2 = dx^2 + dy^2 + dz^2,$$

which can be described by the metric

$$g_{ij} = \begin{pmatrix} 1 & 0 & 0 \\ 0 & 1 & 0 \\ 0 & 0 & 1 \end{pmatrix}.$$

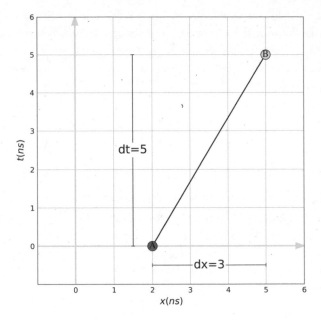

FIGURE 2.4. The displacement between two nearby points in spacetime. In this case: $dx = 3$ ns and $dt = 5$ ns.

There is a magic to this metric. It stays the same even if you move the origin (displacement) or rotate the axes, as we showed in the previous chapter. No matter what you do, the distances between two points remain the same, even when the individual components of displacement change.

Can we find a similar relation in spacetime? Of course we can! We start by writing down a displacement:

$$dx^\mu = \begin{pmatrix} dt \\ dx \\ dy \\ dz \end{pmatrix},$$
(2.4)

which can be visualized on a spacetime diagram as in figure 2.4.

Since we've already insisted that space and time are *equivalent* in a sense, you might suppose (wrongly) that the invariant relation in spacetime is:

$$\text{distance}^2 = dt^2 + dx^2 + dy^2 + dz^2$$

(I crossed it out to make sure you don't accidentally write this down as the correct relation.)

As an alternative to the relation above, we're going to guess a definition of the **interval** between two events as

$$ds^2 \equiv dt^2 - dx^2 - dy^2 - dz^2, \qquad (2.5)$$

or, written as tensor relation

$$ds^2 = g_{\mu\nu} dx^\mu dx^\nu, \qquad (2.6)$$

where $g_{\mu\nu}$ is known as the **Minkowski metric** and takes the form

$$g_{\mu\nu} = \begin{pmatrix} 1 & 0 & 0 & 0 \\ 0 & -1 & 0 & 0 \\ 0 & 0 & -1 & 0 \\ 0 & 0 & 0 & -1 \end{pmatrix}. \qquad (2.7)$$

The convention of making the time term positive and the spatial terms negative isn't universal. Some texts do it the other way around. However, our notation will turn out to be especially useful in particle physics applications, so we'll stick with it.

We can write the four-dimensional dot product via

$$ds^2 = d\mathbf{x} \cdot d\mathbf{x}.$$

This is algebraically identical to equation (2.6) but doesn't explicitly list the indices.

Why is the interval important? Consider a light beam propagating in the $+x$ direction. For every interval of time dt, it will traverse an identical spatial distance, dx. Or, in other words,

$$ds^2 = dt^2 - dx^2 = 0.$$

As we've seen experimentally, the speed of light is the same for all reference frames, which means that if an interval is zero in one frame of reference, it's zero in all. We haven't quite shown that the the Minkowski metric is generally invariant, but so far, it seems to work for the special case of light beams.

In our particular case, the interval squared is positive. This is known as a **timelike interval**. As we've already seen, if the interval is zero, it means space = time, so the interval is **null** or **lightlike**. We'll talk about what happens when the square interval is negative—so-called **spacelike intervals**—in due course.

In ordinary 3-space, when we take the dot product of a vector with itself, we get the magnitude squared. Provided we've defined our metric well, we can rotate or transform our coordinate system however we like and, as we've seen, it won't change the inferred magnitude of the vector.

Example: Interval and Speed

Consider two events depicted by the spacetime diagram below.

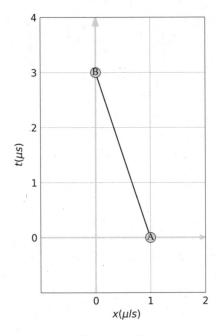

Compute both the interval between the two events and the velocity of a signal propagating between the two.

Solution
We first note that

$$\Delta x = -1; \quad \Delta t = 3,$$

so

$$\Delta s^2 = \Delta t^2 - \Delta x^2 = 8,$$

yielding an interval of

$$\Delta s = \sqrt{8},$$

which is positive, and thus timelike separated. The velocity is

$$u = \frac{\Delta x}{\Delta t} = -1/3 \ .$$

So far, I'm saying, but not yet proving, that the Minkowski metric allows us to do the same in spacetime. The interval generated for a light beam is zero, but we haven't yet shown that the Minkowski metric is true in general.

This also gives us an opportunity to introduce yet another tensor rule:

5. The metric allows you to raise or lower an index.

Behold the equation

$$U_\nu = g_{\mu\nu} U^\mu. \tag{2.8}$$

This equation satisfies the summation rule and the matching indices rule, but in the case of the metric, we've got something special. It allows us to take a vector, for instance, and turn it into a 1-form. In the case of the vector above,

$$U_\nu = \begin{pmatrix} 5/3 & 0 & -4/3 & 0 \end{pmatrix}.$$

The space terms get minus signs.

Example: Lowering Indices

Consider a relatively sparse rank-2 tensor:

$$T^{\mu\nu} = \begin{pmatrix} 2 & 0 & 0 & 0.5 \\ 0 & 0 & 0 & 0 \\ 0 & 0 & 0 & 0 \\ 0.8 & 0 & 0 & 1 \end{pmatrix},$$

where the first index identifies the row and the second, the column.
What is the lowered version of the tensor $T_{\alpha\beta}$?

Solution
We do this in two steps, one for each index. The order doesn't matter, but first, we'll lower the second index. For the spacelike terms ($\nu = 3$),

we'll acquire a minus sign:

$$T^{\mu}_{\ \beta} = g_{\nu\beta}T^{\mu\nu}$$

$$= \begin{pmatrix} 2 & 0 & 0 & -0.5 \\ 0 & 0 & 0 & 0 \\ 0 & 0 & 0 & 0 \\ 0.8 & 0 & 0 & -1 \end{pmatrix}.$$

Now do the same for the first index. Again, for $\mu = 3$, we'll get a minus sign, which will flip the T_{33} back to the original:

$$T_{\alpha\beta} = g_{\alpha\mu}T^{\mu}_{\ \beta}$$

$$= \begin{pmatrix} 2 & 0 & 0 & -0.5 \\ 0 & 0 & 0 & 0 \\ 0 & 0 & 0 & 0 \\ -0.8 & 0 & 0 & 1 \end{pmatrix}.$$

2.3.3 WHERE THE MINKOWSKI METRIC COMES FROM

The speed of light is *the* measurement tool in relativity, and we're going to exploit its constancy to derive just about everything. For instance, consider a **light clock** inside a stationary spaceship (figure 2.5).

A photon or light pulse is shot from the floor to the ceiling of a ship, reflected downward, and reabsorbed at the floor. The total travel time is $dt = 2h$, but since the photon is detected right next to where it was emitted,

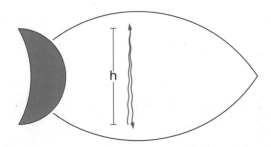

FIGURE 2.5. The light clock. A photon is fired from floor to ceiling, reflected downward, and detected. The total distance traveled is $2h$, and thus (in MKS units), the time interval for the trip is fixed as $2h/c$, or $dt = 2h$ in natural units.

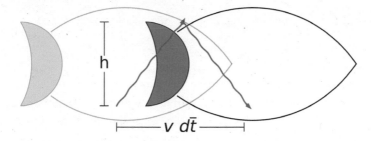

FIGURE 2.6. The same as in figure 2.5, but with the ship moving to the right at speed v.

the spatial seperation between the emission and absorption is zero. Thus, the displacement between the two is

$$dx^\alpha = \begin{pmatrix} 2h \\ 0 \\ 0 \\ 0 \end{pmatrix},$$

yielding an interval squared of $ds^2 = 4h^2$.

But now imagine that the light clock is situated within a ship (conveniently made of glass) traveling at speed v with respect to observers around them (figure 2.6).

In this new frame (the one outside the rocket), the duration of the light clock is potentially unknown; we will simply call it $d\bar{t}$. And thus, the new displacement between the "ticks" of the light clock is

$$dx^{\bar\alpha} = \begin{pmatrix} d\bar{t} \\ d\bar{x} \\ 0 \\ 0 \end{pmatrix}.$$

But we have one more trick up our sleeves: the constant speed of light. The light path may be divided up into two legs (up and down), and each can be determined by the Pythagorean theorem:

$$\text{distance}^2 = 4\left(h^2 + \left[\frac{d\bar{x}}{2} \right]^2 \right).$$

But we also know that because we're talking about a light beam, distance $= d\bar{t}$. Thus,

$$d\bar{t}^2 = 4h^2 + d\bar{x}^2.$$

But recall $4h^2 = ds^2$, our original interval. Substituting, we get

$$ds^2 = d\bar{t}^2 - d\bar{x}^2.$$

This is the Minkowski metric in a new, moving frame!

What we've done is called a **boost**, and it's simply a way of computing properties in one frame that is moving at a constant speed with respect to another. And we've shown that under boosts, the interval remains conserved—not just for $ds^2 = 0$, but for *any* interval.

2.4   Einstein's Postulates

We have described the conditions of relativity, but haven't really set up the rules. While we tend to associate relativity with Einstein, the first formulation came from Galileo in his *Dialogue Concerning Two New Sciences* [19]:

> Shut yourself up with some friend in the main cabin below decks on some large ship, and have with you there some flies, butterflies, and other small flying animals. Have a large bowl of water with some fish in it; hang up a bottle that empties drop by drop into a wide vessel beneath it. With the ship standing still, observe carefully how the little animals fly with equal speed to all sides of the cabin. The fish swim indifferently in all directions ... When you have observed all these things carefully (though there is no doubt that when the ship is standing still everything must happen in this way), have the ship proceed with any speed you like, so long as the motion is uniform and not fluctuating this way and that. You will discover not the least change in all the effects named, nor could you tell from any of them whether the ship was moving or standing still.

This is the foundational idea in what we now call Newtonian mechanics: you can't tell if you're moving or standing still. You can only observe *relative* motion. This will form our basis for relativity as well. The **principle of relativity** [9, 13] says:

> 1. The laws according to which the nature of physical systems alter are independent of the manner in which these changes are referred to two co-ordinate systems which have a uniform translatory motion relative to each other.

Or, more plainly, *physical laws are the same whether you are in a moving or stationary reference frame.*

So far, Einstein and Galileo are saying almost exactly the same thing. Since Galileo formed the basis for Newtonian physics, we can even infer something from this. Let's imagine a three-dimensional particle trajectory as seen in two frames:

$$x^{\bar{i}}(t) = x^i(t) + v^i t,$$

so

$$\frac{dx^{\bar{i}}}{dt} \equiv u^{\bar{i}}$$
$$= u^i + v^i, \tag{2.9}$$

and

$$a^{\bar{i}} = a^i.$$

This is another way of saying that, in Newtonian mechanics, accelerations are frame independent. I used u^i to refer to the velocity of a particle, while v^i was used for the relative velocity of a frame. This will be our convention throughout.

In one dimension (at low speeds), this means

$$d\bar{x} = dx + vdt.$$

In other words, separations in different frames are *not* the same.

Note that this is the transformation for *Galilean* relativity, not *special* relativity. Why? Because it allows particles to go faster (or slower) than the speed of light. That brings us to Einstein's second postulate:

2. Every ray of light moves in the "stationary co-ordinate system" with the same velocity c, the velocity being independent of the condition whether this ray of light is emitted by a body at rest or in motion. Therefore

$$\text{velocity} = \frac{\text{path of light}}{\text{interval of time}}.$$

Or, again in plain English, *light travels at the same speed for all inertial observers*.

Remember that the interval ($ds^2 = dt^2 - dl^2$) gives 0 for light, so if it really is the same for everyone, then "interval is invariant" automatically guarantees "speed of light is the same."

Galileo is able to satisfy the first principle but not the second. We'll devote the next chapter to reconciling this issue.

Looking Forward

In this chapter, we developed the foundations of spacetime and found that, in a lot of ways, the mathematics of Euclidean space are not so different from those of Minkowski spacetime (but for those pernicious minus signs in the metric!). Given that the speed of light is the same in all reference frames, there's something odd about the workings of space and time that are different from what Galileo and Newton assumed. In the next chapter, we'll quantify all that and find out why it is that time does not, in fact, flow at the same rate to all observers.

2.5 Problems

1. Compute the following in natural units of time:
   (a) The distance to the Sun
   (b) The distance to Earth's moon
   (c) 1 foot
2. Compute the following conversions in natural units:
   (a) 50 kg in units of energy
   (b) The mass of an electron expressed in inverse distance
   (c) The mass of a helium nucleus expressed in inverse time
   (d) 1 nm expressed in inverse energy
3. Events A, B, and C are plotted on the spacetime diagram below:

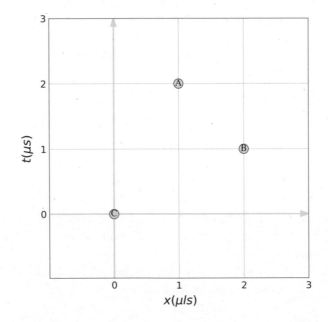

(a) What is the interval squared between i) A and B; ii) B and C; iii) A and C?

(b) Which of these intervals are i) spacelike separated? ii) timelike separated? iii) lightlike separated?

(c) Suppose a projectile was sent from C to A. At what speed does it travel?

4. Consider the following events in $1+1$ dimensional spacetime (all units in seconds):

$$x_A^\mu = \begin{pmatrix} 1 \\ 2 \end{pmatrix}; \quad x_B^\mu = \begin{pmatrix} -2 \\ 5 \end{pmatrix}; \quad x_C^\mu = \begin{pmatrix} 3 \\ 3 \end{pmatrix}.$$

(a) Plot all three events on a spacetime diagram.

(b) For all three pairs of points (e.g., AB), compute the interval squared using the Minkowski metric.

(c) For all three pairs, indicate whether the separation is spacelike, timelike, or lightlike. For any timelike separations, indicate which event occurred first.

5. A star orbits a black hole at a radius of 4 light-seconds and travels in a circular orbit at a speed of 0.4 in the xy-plane. Draw a spacetime diagram representing the x-coordinate position of the star. Be sure to show at least one full orbit.

6. Draw a spacetime diagram with both axes equally spaced in units of seconds (or light-seconds) and include the following features:

(a) A worldline for stationary observer A at $x = -1$ and another for stationary observer B at $x = 1$.

(b) At $t = 0$, A sends a radio message to B.

(c) When the message is received by B, they wait 1 second and then send a response back by a particle beam traveling at $0.5c$.

(d) Create an event representing A's receipt of the return message, and calculate the amount of time for the total exchange.

FURTHER READINGS

- E. A. Abbott, *Flatland: A Romance of Many Dimensions* (Seeley & Co., 1884). While this science fiction classic is deeply problematic in terms of sexism and classism (the number of interior angles correlate with intelligence in this world, and women are denoted as straight lines), it remains a useful popular read from the perspective of viewing the worlds of different numbers of dimensions.

- S. M. Carroll, *From Eternity to Here: The Quest for the Ultimate Theory of Time* (Dutton, 2010). A well-written popular work on the history and nature of the arrow of time in which the author describes some of his own work on the subject.

- A. Einstein, *Relativity: The Special and General Theory*, introduction by Roger Penrose and with commentary by Robert Geroch (Pi Press, 2005). Einstein's original popular descriptions of relativity are very readable, and the interested student will learn a lot. The individual works are largely in the public domain at

this point, so there are many reprints. This is a particularly nice collection with
very useful commentary.

- D. M. Goldberg, *The Universe in the Rearview Mirror: How Hidden Symmetries Shape Reality* (Dutton, 2013). This is a popular approach that I wrote regarding symmetries in the physical universe. It contains an extensive conceptual discussion of symmetries generally and special relativity in particular.

- R. Le Poidevin, *Travels in Four Dimensions: The Enigmas of Space and Time* (Oxford University Press, 2003). While we'll explore paradoxes in special relativity in Chapter 4 of this book, Le Poidevin's series of essays focuses on mysteries related to the flow of time or treating time as a dimension more generally.

- C. Rovelli, *The Order of Time* (Riverhead Books, 2018). A bit more philosophical than the typical recommendation, this book is an outstanding overview of the nature of spacetime and the arrow of time.

Lorentz Transforms

H. A. Lorentz (1853–1928) developed a formalism for converting between one boosted frame and another. Although Lorentz developed his theory under the assumption that the æther was compressed along the direction of motion, his work was foundational for the spacetime development of special relativity.

3.1 Conic Sections

We went through a lot of effort to justify the Minkowski metric in the previous chapter, and you might be wondering why. The central premise of the first few chapters is that you started this book with an intuitive understanding about how flat space works, and that we'd be able to exploit that intuition to help you understand the nuances of spacetime. As a quick reminder, let's look at the metric for two-dimensional Euclidean space and see what it tells us:

$$dl^2 = dx^2 + dy^2.$$

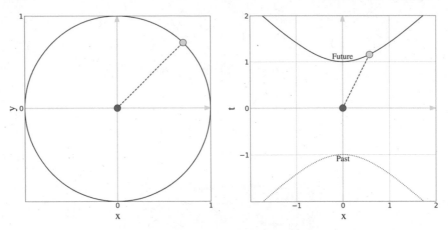

FIGURE 3.1. Left: A visualization of the Euclidean metric. All points have a $dl^2 = 1$ from the origin. Rotating the axes leave the metric unchanged. Right: All points have a fixed spacetime interval of $ds^2 = dt^2 - dx^2 = 1$ from the origin. The upper branch of the hyperbola shows those in the future, and the lower branch shows the past.

The idea of the metric is that if you rotate the coordinates, distances between points remain the same, so if you look at the the right-hand side of the relation, you get $dx^2 + dy^2 = $ constant, which is the equation for a circle (figure 3.1, left panel).

In the last chapter, we found that the fixed speed of light—Einstein's first postulate of relativity—had some unanticipated consequences. For one (in 1+1 dimensions), the interval is

$$ds^2 = dt^2 - dx^2.$$

What can you do to keep this interval from changing? As with rotations (wherein you simultaneously change x and y), with a boosted frame,* both time and space are going to need to simultaneously change.

We don't know all the details yet, but suppose we have an event at the origin

$$x^\mu = \begin{pmatrix} t \\ x \end{pmatrix} = \begin{pmatrix} 0 \\ 0 \end{pmatrix},$$

---

*Though we've used boosts before, it's a technical sounding term that is conceptually very simple. A frame of reference is boosted with respect to another if the observers are moving with respect to one another. It's illustrated by the effect of measuring physics within a moving train as compared with measuring effects seen outside the train.

within some reference frame. A second later (in the observer's own reference frame), the coordinates are $\left(\begin{smallmatrix} 1 \\ 0 \end{smallmatrix}\right)$, with a total interval squared of $ds^2 = 1$.

One of the big goals in relativity is not just to show that "all things are relative," (which, sure, they are, but you don't need a physics class to tell you that) but rather to find quantities that are the same for all observers. The speed of light is the big one. But there are others that are a bit more subtle.

There's a technical name to the time that passes on a clock that travels with an observer: the **proper time**, which is frequently given the variable label τ, but this only works for timelike separated events. There's no proper time for a lightlike interval—there's no rest frame for a photon—or spacelike intervals. Two spacelike separated events would require a particle to travel faster than the speed of light to traverse between them.

Different reference frames will see things differently. Again, we don't know exactly how to convert to a specific frame (any more than we knew how to rotate before deriving the rotation matrix (eq. 1.9). However, we *do* know that

$$dt^2 - dx^2 = \text{constant}.$$

As with two-dimensional Euclidean space, which had a circular symmetry, 1+1 Minkowski space has the symmetry of a different conic section, the hyperbola.* This is shown in in the right panel of figure 3.1.

Looking at the spacetime hyperbola should immediately make you nervous about your spacetime intuition. In ordinary Euclidean space, what we call x or y or z is arbitrary, but surely *time* should be fixed. But as we quickly see in figure 3.1, the time interval between two events very much depends upon the frame of reference. Time isn't constant after all!

But there are even more nonintuitive factors related to all of this. Because all inertial observers have the right to think of themselves as standing still, and because $ds^2 = 1$ everywhere along the upper hyperbola, to an observer moving between the origin and any point on that locus, their watch would only tick forward 1 second, even though more time passes to the outside observer.

Einstein's second postulate—the equivalence of all inertial frames— provides the foundation for developing the Lorentz transforms—the equations that will provide the relativity to relativity. These transforms will tell

---

*It may take a fair amount of brain stretching to think back to conic sections. These are the figures traced by slicing through a pair of cones; besides circles and hyperbolas, they also include ellipses and parabolas.

us how to convert from one frame of reference to another. To derive those transforms, we'll need to go to space.

Example: Proper Time

A spaceship captain makes an expedition to Barnard's star, 6 light years from Earth. According to Earth's clocks, the trip takes 10 years ($v = 0.6c$). How much time passes for the captain?

Solution
The proper time squared is

$$d\tau^2 = (10\,\text{yr})^2 - (6\,\text{yr})^2 = 64\,\text{yr}^2,$$

or, the proper time is $\tau = 8\,\text{yr}$.

3.2   Time Dilation

3.2.1   ROCKET SCIENCE

Imagine yourself as the victim of a very complicated and intricate spacetime prank. In the middle of the night, your labmates carefully move your bed into a vast spaceship. You awake to find yourself in a very dark, barren region of space where there are no external indications as to your speed. The engines, at any rate, are turned off, so the ship is neither accelerating nor decelerating.* Without any reason for assuming otherwise, you may as well unilaterally decide that the ship is at rest and measure everything with respect to markings in the interior of the ship. This frame of reference is shown in figure 3.2. In this spacetime diagram, we've plotted two events corresponding to the ticking of a clock, or $dt = d\tau$, because the coordinate of time on the ship *is* the proper time. This is, in fact, the spacetime diagram version of our light clock from earlier.

Though the ship has its own reference frame, we can imagine that, unknown to you, there is another observer outside the ship, and from the alien's perspective, the ship is moving swiftly to the right at speed v, as shown in figure 3.3. To distinguish between the two perspectives, I've used barred coordinates $\bar{t}$ and $\bar{x}$ to denote the alien's point of view.

---

*Even your terrestrial experiences would inform you that it's easy to tell if you're accelerating because you would be thrown to the back of the ship.

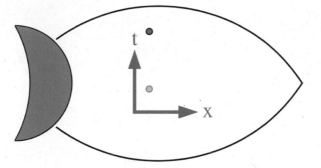

FIGURE 3.2. The frame of reference as seen within a rocket ship. We've marked off two points in its internal spacetime diagram. They are separated in time rather than in space, so these two events might represent the ticking of a clock. A bit confusingly, perhaps, "up" in this diagram refers to the future.

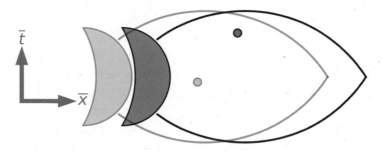

FIGURE 3.3. The same ticking of a clock seen in figure 3.2 but from a different perspective. As drawn, the ship is moving at $v = 0.8$ compared to an observing alien.

While the ticks of the clock occur at the same point in space in the unbarred reference frame, they are in different positions in the barred frame. As seen within the ship,

$$ds^2 = dt^2 - dx^2 = d\tau^2.$$

This is the situation depicted in the right panel of figure 3.1. Once again, this is a timelike interval. Regardless of the reference frame, all observers will agree that tick 1 comes before tick 2.

You should internalize the idea that, in different frames of reference, spacetime trajectories that are longer on the diagram (at fixed t duration) are going to be *shorter* proper times.

We can switch to the outside observer—the barred frame. We do know a little about it. Making no assumptions about how long the clock appears to tick from outside the ship, we know that between ticks, the ship moves a distance of

$$d\bar{x} = v d\bar{t}.$$

Thus, the interval is

$$d\bar{s}^2 = d\bar{t}^2 - d\bar{x}^2 = d\bar{t}^2 - v^2 d\bar{t}^2.$$

But wait! The interval needs to be the same both inside the ship and in any other inertial reference frame. Thus, for a stationary clock,

$$d\tau^2 = (1 - v^2) d\bar{t}^2$$

$$d\bar{t} = \frac{d\tau}{\sqrt{1 - v^2}}. \tag{3.1}$$

We can plot the rocket's observational frame in figure 3.2. The clocks tick *slower* in a moving reference frame. This means *every* measure of time. A moving observer will age slower, clocks will run slower, particles will decay slower (more on that later), and so on. The effect is real, and it is measurable.

We'll come up with a general theory of frames of reference in the next section, but for now, we're going to think a little more about moving clocks.

3.2.2 THE GAMMA FACTOR

In equation 3.1 we saw that, as seen from the outside, your moving clock seemed to be slowed by a factor of

$$\gamma \equiv \frac{1}{\sqrt{1 - v^2}}. \tag{3.2}$$

This **gamma factor** (or Lorentz factor) shows up just about everywhere in special relativity, and it's important enough that it's worth taking a brief digression.

For moving clocks, it corresponds to **time dilation** (γ), the amount by which the clock appears to be slowed. Mechanically, it's straightforward to compute γ, but I've put a few common values in table 3.1 to get you started.

As a practical matter, normal terrestrial conditions are almost always consistent with $\gamma \approx 1$. Consider the orbital speed of Earth around the Sun, $v = 30,000 \, \text{m/s} = 10^{-4}$. The corresponding γ factor is 1.0000005, a value that doesn't differ meaningfully from 1. Put another way, under normal terrestrial conditions, time doesn't differ meaningfully between frames.

TABLE 3.1 Some common gamma factors.

v	γ
0	1
0.1	1.005
0.6	1.25
0.8	1.666
0.9	2.29
0.99	7.09
0.999	22.4

In addition, the general form of computing γ tells us a couple of things right away.

1. γ is always greater than or equal to 1. It's approximately 1 at nonrelativistic speeds, as we've seen. Indeed, in the limit of $v \ll 1$, we can use a Taylor series expansion of a binomial,

$$\lim_{x \to 0}(1+x)^n \approx 1 + nx, \tag{3.3}$$

to yield

$$\lim_{v \to 0} \frac{1}{\sqrt{1-v^2}} \approx 1 + \frac{1}{2}v^2, \tag{3.4}$$

a relation that we'll refer back to frequently.
2. As $v \to 1$ (approaches the speed of light), γ gets larger and larger. It approaches infinity as v goes to 1.

A general relation of gamma factors and speeds is plotted in figure 3.4. Because gamma factors will show up in almost every aspect of relativity, you should practice converting from speed to γ now, and I've put a number of exercises at the end of the chapter to help you do so.

You will also, on occasion, want to go from γ to v, as they are one-to-one functions of one another. The algebra is straightforward, but a little awkward:

$$\gamma = \frac{1}{\sqrt{1-v^2}}$$

$$\frac{1}{\gamma^2} = 1 - v^2,$$

so

$$v = \sqrt{1 - \frac{1}{\gamma^2}}, \tag{3.5}$$

which makes the asymptotic limits of $v \to 0$ as $\gamma \to 1$ and $v \to 1$ as $\gamma \to \infty$ even more obvious.

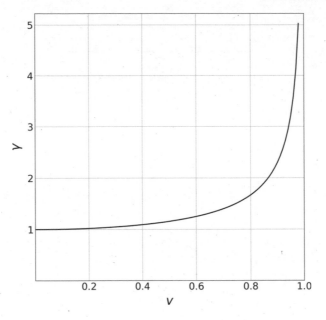

FIGURE 3.4. The γ factor as a function of v.

3.2.3 TIME DILATION IS FOR REAL

Now is as good a time as any to answer the question that might be on your mind: Is any of this real, or is it just a mathematical exercise?

Oh, it is absolutely real.

In 1941, Bruno Rossi and David Hall [39] found that charged elementary particles known as **muons** created in the upper atmosphere survived to the surface of Earth (figure 3.5) despite the fact that their short mean lifetime of 2.4 μs should have destroyed them before completing their journey. The explanation was that because the muons were traveling at relativistic speeds, their internal "clocks" were slowed relative to Earth and, thus, their effective decay time was effectively increased.

Example: Muon Decay

Muons are created approximately 10 km above the surface of the Earth. What is the typical speed of atmospheric muons if 5 percent of them reach the surface of the Earth?

Solution

Even traveling at the speed of light, it should take muons 33 μs to reach the surface from the upper atmosphere, many times their mean lifetime. If 5 percent of muons survive, then, in the frame of the muons, the trip to Earth takes only

$$e^{-t/\tau} = 0.05; \quad t = 3\tau = 7.2 \, \mu s.$$

Time must be dilated by a factor of $\gamma = \frac{33 \, \mu s}{7.2 \, \mu s} = 4.58$.
    Inverting the γ relation,

$$v = \sqrt{1 - \frac{1}{\gamma^2}} = 0.976.$$

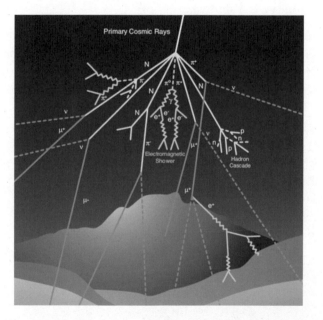

FIGURE 3.5. The chain of particles and decays in the upper atmosphere is fairly complicated, but in the absence of relativistic time dilation, *none* of the muons created in the atmosphere should make it to the ground—but they *do*! Courtesy: CERN.

Since that time, we haven't had to rely solely on fundamental particles. We're able to compare high-precision clocks in high-speed satellites, the International Space Station, and even ordinary airplanes. The effect is

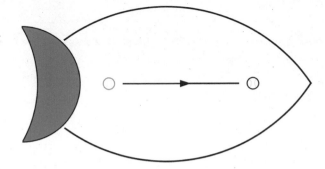

FIGURE 3.6. A thrown ball as seen from the interior of a "stationary" spaceship.

small—a 10-hour flight will produce a time dilation of only a few 10s of nanoseconds—but measurable.

3.3   The Galilean Transforms

We saw earlier that Galileo thought about the relative motion of a ship on a smooth sea as his basis for considering **Galilean relativity** (space is relative but time is not). We're going to generalize a bit and pretend that Galileo had access to spaceships and imagine the motion of a spaceship at a constant speed.

Throughout this discussion, I'm frequently going to switch between barred and unbarred frames. For convenience, I'll generally write the unbarred frame as being the one *inside* the spaceship, as depicted in figure 3.2, while the barred frame is the view from the outside (figure 3.3).

The Galilean approach to translating between these frames is more or less what you would have guessed intuitively. Suppose you throw a baseball from the back of the ship to the front, as in figure 3.6. From the perspective inside the ship, the ball traverses a "horizontal" distance of Δx, and the ball takes time Δt. Thus, the speed of the ball is

$$u = \frac{\Delta x}{\Delta t},$$

where the unusual aspect so far is that we're using u (rather than v, which will represent the speed of a ship or a frame from here on out) to represent the speed of the ball.

From outside the ship, things are slightly different. While Galileo thought that the clock would tick at the same rate, he was well aware that the ball would appear to travel further than seen from the inside (figure 3.7).

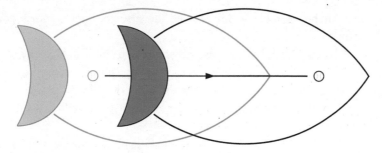

FIGURE 3.7. A thrown ball as seen from the outside of a moving spaceship.

As seen from the outside,

$$\Delta \bar{t} = \Delta t; \quad \Delta \bar{x} = \Delta x + v\Delta t,$$

because the front of the ship moves considerably during the flight of the ball. Combining the two perspectives, Galileo estimates that the barred frame would see the ball moving at

$$\begin{aligned} \bar{u} &= \frac{\Delta \bar{x}}{\Delta \bar{t}} \\ &= \frac{\Delta x + v\Delta t}{\Delta t} \\ &= u + v, \end{aligned}$$

exactly as intuition would have suggested. The velocities are just added to each other.

But this creates a problem, because if you were to fire a laser inside a spaceship, then outside the ship, the light would seem to travel at $c + v > c$, faster than the speed of light. And because we've already established experimentally that light travels at the same speed to all observers, this is clearly wrong.

3.4 The Lorentz Transforms

We've focused on translating very specific situations (a set of light pulses emitted simultaneously in opposite directions, for instance) when translating between one frame and another. The generalized formulas for doing this are known as the **Lorentz transforms.**

Like so much in physics, the Lorentz transforms were named after the *last* person to propose them, rather than the first. A version of the equations was originally discovered by Woldemar Voigt [44] in 1887 (the same year, you may recall, that Michelson and Morley showed the speed of light is independent of direction).

H. A. Lorentz [25, 26] based his conjectures on the idea that the æther was real. He supposed that, as the measurement apparatuses moved through the æther, they were smooshed along the direction of motion. He was mostly concerned about the implications for electromagnetism, which we haven't touched on yet (but we will in Chapter 7).

It wasn't until 1905 that Einstein seriously explored the possibility that time really is dilated within different frames of reference. Einstein's working assumption was that space and time really are different in different inertial reference frames—that is to say, frames that are boosted with respect to one another.

We're going to do exactly the same thing that we did when we talked about rotational invariance (eq. 1.9), which is to assume a linear coordinate transformation:

$$dx^{\bar{\mu}} = \Lambda^{\bar{\mu}}_{\ \mu} dx^{\mu}.$$

Expanding this out in $1 + 1$ dimensions,

$$d\bar{t} = \Lambda^{\bar{t}}_{\ t} dt + \Lambda^{\bar{t}}_{\ x} dx$$

and

$$d\bar{x} = \Lambda^{\bar{x}}_{\ t} dt + \Lambda^{\bar{x}}_{\ x} dx.$$

When we evaluated time dilation in a moving clock (leading to eq. 3.1), we set up the clock to move along with the spaceship, so $dx = 0$. Thus, we haven't yet computed $\Lambda^{\bar{t}}_{\ x}$ or $\Lambda^{\bar{x}}_{\ x}$. That's our missing step. But we can make an educated guess about their functional forms:

$$d\bar{t} = \gamma dt + f(v) dx.$$

and

$$d\bar{x} = v\gamma dt + g(v) dx.$$

We can just plug that into the Minkowski metric:

$$d\bar{t}^2 - d\bar{x}^2 = \gamma^2 dt^2 + f(v)^2 dx^2 + 2\gamma f(v) dt dx - v^2\gamma^2 dt^2 - g(v)^2 dx^2 - 2g(v)v\gamma dx dt$$
$$= \gamma^2(1 - v^2)dt^2 + 2\gamma\left[f(v) - vg(v)\right] dx dt - \left[g(v)^2 - f(v)^2\right] dx^2.$$

The expression on the right looks really ugly, but because the entire goal of this work is that the metric needs to be the same in all frames,

$$d\bar{t}^2 - d\bar{x}^2 = dt^2 - dx^2,$$

then the pre-factors in front of dt^2, dx^2 and $dx\,dt$ simplify dramatically:

$$\gamma^2(1-v^2) = 1$$
$$2\gamma\left[f(v) - vg(v)\right] = 0$$
$$-\left[g(v)^2 - f(v)^2\right] = -1.$$

The first of these is already true. The second gives

$$f(v) = vg(v),$$

and plugging that result into the third yields

$$g(v)^2(1-v^2) = 1,$$

which means that the aptly named $g(v) = \gamma$ and $f(v) = v\gamma$. Putting it all together results in the following very pleasingly symmetric forms:

$$d\bar{x} = \gamma dx + v\gamma dt \qquad\qquad (3.6)$$

$$d\bar{t} = \gamma dt + v\gamma dx. \qquad\qquad (3.7)$$

Or, as a matrix:

$$\Lambda^{\bar{\mu}}{}_{\mu} = \begin{pmatrix} \gamma & v\gamma \\ v\gamma & \gamma \end{pmatrix}, \qquad\qquad (3.8)$$

which has the nice properties that

$$\det\Lambda = (1-v^2)\gamma^2 = 1.$$

The inverse is computed by reversing the velocity:

$$\Lambda^{\mu}{}_{\bar{\mu}} = \begin{pmatrix} \gamma & -v\gamma \\ -v\gamma & \gamma \end{pmatrix}, \qquad\qquad (3.9)$$

which allows us to transform from the barred frame to the unbarred.

Example: Spaceship Boost

An astronaut in a spaceship fires a small projectile from a relativistic gun. It travels $\Delta x = 1\mu\text{ls}$ in a period of $\Delta t = 2\mu\text{s}$, moving at half the speed of light.

As seen from the outside, the ship is traveling at $v = 0.6$. How far does the projectile travel, and how long does it take as seen from outside the ship?

Solution

As with most problems of this sort, we begin by computing the γ factor:

$$\gamma = \frac{1}{\sqrt{1-(0.6)^2}} = 1.25.$$

Thus, the transformation matrix is

$$\Lambda^{\bar{\mu}}{}_{\mu} = \begin{pmatrix} \gamma & v\gamma \\ v\gamma & \gamma \end{pmatrix} = \begin{pmatrix} 5/4 & 3/4 \\ 3/4 & 5/4 \end{pmatrix}.$$

We can take the determinant of this particular transformation matrix, and indeed, it is $25/16 - 9/16 = 1$, as promised.

Finally, to compute the spacetime displacement in the barred (external) frame, we get

$$\begin{pmatrix} \Delta\bar{t} \\ \Delta\bar{x} \end{pmatrix} = \begin{pmatrix} 5/4 & 3/4 \\ 3/4 & 5/4 \end{pmatrix} \begin{pmatrix} 2 \\ 1 \end{pmatrix} = \begin{pmatrix} 13/4 \\ 11/4 \end{pmatrix}.$$

You will note that, as seen from outside the ship,

$$\bar{u} = \frac{\Delta\bar{x}}{\Delta\bar{t}} = \frac{11}{13} \approx 0.85.$$

This sum is less than Galileo would have guessed by simply adding the speed of the ship and the projectile together ($1.1c$). And, notably, it's less than the speed of light.

Let's consider what happens in the limit of $v \to 0$ (nonrelativistic speeds). In that case, $\gamma = 1$, and we're left with

$$d\bar{x} = dx + vdt,$$

which is exactly what Galileo gives, and

$$d\bar{t} = dt + vdx,$$

which is not. But that's only because the second term is very small, $v \ll 1$ and $dx \ll dt$, for all reasonable nonrelativistic conditions, since (small) × (small) = tiny.

3.4.1 VELOCITY ADDITION

In Galilean relativity, we were left with a conundrum: If you turn on the headlights of a spaceship, they seem to travel faster than light to an outside observer, but that's wrong, experimentally.

Example: Velocity Addition

Imagine your ship is moving at 0.5 and you send a relativistic particle at $u = 0.5$. How fast will it appear to an observer outside the ship?

Solution

$$\bar{u} = \frac{u + v}{1 + uv},$$

where, in this case, $u = v = 0.5$. Plugging in, we get

$$\bar{u} = \frac{0.5 + 0.5}{1 + (0.5)(0.5)} = 0.8.$$

Half the speed of light plus half the speed of light is somehow only 80 percent the speed of light. This is not the last mystery of this sort that we will encounter.

So how do velocities add to each other? We just apply the Lorentz transforms:

$$\bar{u} = \frac{d\bar{x}}{d\bar{t}}$$

$$= \frac{\gamma dx + v\gamma dt}{\gamma dt + v\gamma dx}$$

$$= \frac{u + v}{1 + uv}. \tag{3.10}$$

· So, for instance, if you are traveling at v and turn on your headlights ($u = 1$), then outside the ship,

$$\bar{u} = \frac{1+v}{1+(1)(v)} = 1.$$

The speed of light is the same to all observers!

If, on the other hand, $u, v \ll 1$ then $\bar{u} \approx u + v$.

3.5   Length Contraction

We've spent a fair amount of time talking about special relativity's implications for time and causality. But as we've seen, the Lorentz transforms affect spatial calculations as well. As with proper time, two spacelike separated events ($ds^2 < 0$) have a **proper length** between them of

$$L_0^2 \equiv -ds^2 = dx^2 + dy^2 + dz^2 - dt^2. \tag{3.11}$$

Or, in words, the proper length is the distance between two events in the frame of reference in which the events are simultaneous.

However, as we will see, transformations in distances aren't *exactly* the mirror image of time. Moving clocks appear to run slow (as seen from the outside). What about moving rulers? Lay a meterstick (or any other object of known length) on the floor and consider the separation at both ends, as in figure 3.8.

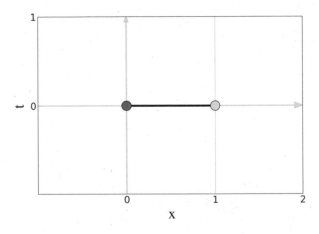

FIGURE 3.8. Two ends of a fixed ruler in its own reference frame. For convenience, I've drawn it as having a length of 1.

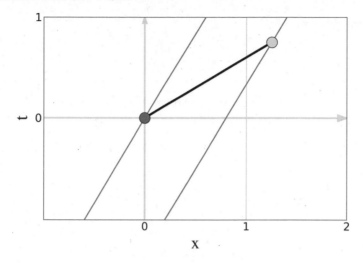

FIGURE 3.9. The same measurements as in figure 3.8 but moving at $v = 0.6$. Note that the measurements of the front and back are no longer simultaneous but, rather, the front is measured at a later time than the back. The dashed lines represent the worldlines of the front and back of the ruler.

Now let's imagine that our meterstick is really the length of a spaceship—the length you'd measure if it were parked next to you. We can come up with all sorts of schemes for measuring it, including standing in the middle of the ship and sending a light pulse forward and backward, bouncing it over each end, and measuring the time the pulses take to return. But rather than say that every time, simply assume that there are marks on the ground and that you're able to measure the length of the ship directly. Most important, you must measure the back and front of the ship *at the same time.*

This may seem obvious. For instance, if I were to measure my kid's height, I would do so by measuring the position of her feet and head at the same time. There are even smartphone apps that do exactly that. Simply point at her feet and then point at her head. If, as kids are prone to do, she jumps up in the air in the moment after I click on her feet, the app would produce a result that's ridiculously wrong. The same would be true if we considered figure 3.8 in a frame in which it was moving, as in figure 3.9.

At first blush, it appears like the ruler is *longer* in the frame in which it's moving. The separation in space, after all, has been increased by a factor of γ. But much as with the example of the jumping kiddo, the real issue is that we're not measuring front and back at the same time. To do that, we need to trace the worldlines of both ends of the ruler, and measure the separation in space only (in the barred frame) for $\Delta \bar{t} = 0$.

That is, we must find

$$\Delta x^{\bar{\mu}} = \begin{pmatrix} 0 \\ L' \end{pmatrix},$$

where, in the unprimed frame, $x_1 = 0$ and $x_2 = L_0$, with unknown Δt.

So, using the Lorentz transforms,

$$\Delta \bar{t} = \gamma \Delta t + v \gamma \Delta x$$

$$= 0$$

$$\Delta t = -v L_0.$$

Thus:

$$\Delta \bar{x} = L' = \gamma \Delta x + v \gamma \Delta t$$

$$= \gamma L_0 + v \gamma (-v L_0)$$

$$= L_0 \gamma (1 - v^2)$$

$$= \frac{L_0}{\gamma}.$$

Surprisingly (or maybe not that surprising if you look at the previous figure or have heard the term before), there is a **length contraction**. The moving ship appears foreshortened along the direction of motion.

Though the derivation of length contraction is more complicated than that of time dilation, it was hypothesized first as a way of explaining the Michelson-Morley experiments. Lorentz's approach in 1892 [25] was developed to explain away the anomalous Michelson-Morley results entirely in terms of spatial terms.* George FitzGerald [15] is often given joint credit for the discovery, despite having only written his conjecture in a short, one-paragraph letter without equations in 1889. While I'll consistently refer to it as "length contraction" here, you'll often see it called "Lorentz contraction" or even "Lorentz-FitzGerald contraction." An oversimplified version of the effect is shown in figure 3.10.

As a practical matter, even were we in a position to observe relativistically moving macroscopic objects without being obliterated by them, the length contraction wouldn't simply appear foreshortened along the direction of motion. In 1959, Roger Penrose [34] and James Terrell [42] independently

---

*It wasn't until a decade later that the more complete form of his eponymous transforms provided the foundation for Einstein's work.

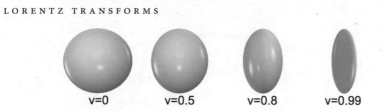

FIGURE 3.10. A sphere moving at relativistic speeds. Because of Doppler shifting light and the potential for apparent "superluminal motion" (see the paradoxes below), it turns out that this is kind of an oversimplification.

noted that, because of the finite speed of light, the rear end of an approaching body would be further retarded than the near side, effectively stretching the image out and canceling the observational effects of length contraction. It would look longer because you'd be observing the front and back at different times (even within your own frame). The **Penrose-Terrell effect** would also result in the images appearing rotated and warped—to the degree that you could see the back side of an approaching cube. A simple ray-tracing simulation is shown in figure 3.11.

We can get a sense of why this might happen by looking at a spacetime diagram in 1+1 dimensions (figure 3.12). As they approach on a direct collision course (somewhat different than the near miss depicted in figure 3.11), it can be shown that a measuring rod is longer than expected by a factor of

$$L' = L_0 \sqrt{\frac{1+v}{1-v}},$$

while the lengths will appear shortened on the way out. As a cube passes directly overhead, for a brief instant, the near side will appear to be contracted by the $1/\gamma$ factor that we computed earlier.

3.6   The 4-Velocity

One of the big goals of relativity is to produce quantities that are the same for all observers. We know from even Euclidean geometry that, while the components of a vector will change from frame to frame, some quantity—the length in 3 space, the interval in 3+1 space—will be invariant.

One scalar quantity that doesn't seem to "work" in relativity is speed. While a change in direction doesn't change the speed of a particle, boosting it certainly does.

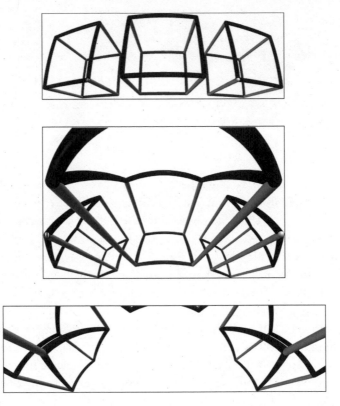

FIGURE 3.11. Three cubes approaching at $v = 0.95$, poised to pass overhead, possibly destroying us all. At a distance, they look relatively cubical, despite the fact that $\gamma \approx 3.2$ at those speeds. As they get closer, the Penrose-Terrell effect becomes more pronounced, with the edges plainly curved, and the cubes (especially the ones on the sides) appearing to be rotated.

Now that we have an *objective* measure of time, we can create a new quantity, the **4-velocity**, which is defined as

$$U^\mu \equiv \frac{dx^\mu}{d\tau}.\tag{3.12}$$

This is obviously only well defined for particles on timelike trajectories.

What is the use of the 4-velocity? Well, consider a 4-velocity dotted with itself:

$$\begin{aligned}
U^\mu U^\nu g_{\mu\nu} &= \left(\frac{dt}{d\tau}\right)^2 - \left(\frac{dx}{d\tau}\right)^2 - \left(\frac{dy}{d\tau}\right)^2 - \left(\frac{dz}{d\tau}\right)^2 \\
&= \frac{1}{d\tau^2}\left[dt^2 - dx^2 - dy^2 - dz^2\right] \\
&= \frac{d\tau^2}{d\tau^2} \\
&= 1.
\end{aligned}$$

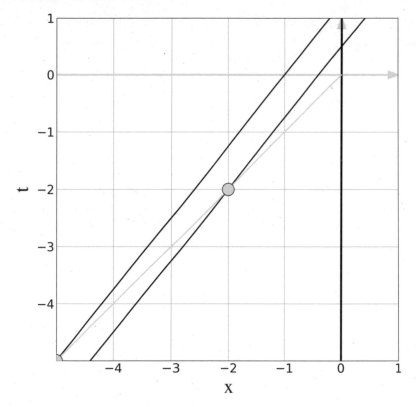

FIGURE 3.12. A rod of length = 1 (in its rest frame) heads toward an observer at the origin at a relative speed of $v = 0.8$. An instantaneous measurement of the box in the observer's frame would yield a length contraction of 0.6. However, in this case, we'd find that the measuring rod is longer than expected.

The magnitude of the 4-velocity is a constant. It's always 1. We love constants in physics. It means, among much else, that if we were to try to measure something like an acceleration we would get the relationship

$$\mathbf{U} \cdot \frac{d\mathbf{U}}{d\tau} = 0.$$

Or, put another way, the 4-acceleration is always perpendicular to the 4-velocity.

But there's more. Consider the quantity U^0, also written as $\frac{dt}{d\tau}$. This is a simple measure of how much time is dilated in the moving frame compared to the reference frame. We have a name for that ratio: γ. Thus, we can use the chain rule for each of the other terms in the 4-velocity to give simplified relations:

$$U^1 = \frac{dx}{d\tau} = \frac{dx}{dt}\frac{dt}{d\tau} = v^1 \gamma.$$

It's the component of the ordinary 3-velocity times the Lorentz factor. Or, in other words,

$$U^\mu = \begin{pmatrix} \gamma \\ v^x\gamma \\ v^y\gamma \\ v^z\gamma \end{pmatrix}.$$

(3.13)

I use a capital U for the 4-velocity of a particle and the lowercase v (as usual) for the 3-velocity. When we write out the values of U^μ, they will include γ factors. But those factors correspond to the speed (v) of the particles within whatever frame you're considering. We can even read off the ordinary speed of the particle pretty easily. Because, $U^1 = v^1\gamma$, and $U^0 = \gamma$, we get a straightforward relation:

$$v^i = \frac{U^i}{U^0}.$$

Example: 4-Velocity

Consider a particle moving with a speed $v = 3/5$ in the $-y$-direction. What is the 4-velocity?

Solution
First, we need to compute γ:

$$\gamma = \frac{1}{\sqrt{1 - v^2}} = 5/4.$$

From here, it is a matter of filling in the blanks:

$$U^\mu = \begin{pmatrix} 5/4 \\ 0 \\ -3/4 \\ 0 \end{pmatrix}.$$

We can then confirm that $\mathbf{U} \cdot \mathbf{U}$ satisfies the normalization condition, as promised:

$$g_{\mu\nu} U^\mu U^\nu = (5/4)^2 - (3/4)^2 = 1,$$

as promised.

The components of the 4-velocity for somebody sitting still in a stationary frame of reference are

$$U^{\mu}_{(rest)} = \begin{pmatrix} 1 \\ 0 \\ 0 \\ 0 \end{pmatrix}.$$

This seems natural enough. They aren't moving through space, but they *are* moving through time at a rate of 1 second per second. This 4-velocity also pretty transparently satisfies the normalization condition, $\mathbf{U} \cdot \mathbf{U} = 1$. To get any arbitrary 4-velocity, we can then boost this stationary 4-velocity in the normal way: $\Lambda^{\bar{\mu}}_{\mu} U^{\mu}_{(rest)}$.

Example: Boosted 4-Velocities

Consider a particle at rest boosted by an amount $v = 4/5$ in the +x direction. What is the corresponding 4-velocity?

Solution

This is almost identical to the previous example, but with a different approach. In this case, $\gamma = 5/3$, so

$$\Lambda^{\bar{\mu}}_{\mu} = \begin{pmatrix} \gamma & v\gamma & 0 & 0 \\ v\gamma & \gamma & 0 & 0 \\ 0 & 0 & 1 & 0 \\ 0 & 0 & 0 & 1 \end{pmatrix} = \begin{pmatrix} 5/3 & 4/3 & 0 & 0 \\ 4/3 & 5/3 & 0 & 0 \\ 0 & 0 & 1 & 0 \\ 0 & 0 & 0 & 1 \end{pmatrix}.$$

And

$$U^{\bar{\mu}} = \Lambda^{\bar{\mu}}_{\mu} U^{\mu}_{(rest)} = \begin{pmatrix} 5/3 & 4/3 & 0 & 0 \\ 4/3 & 5/3 & 0 & 0 \\ 0 & 0 & 1 & 0 \\ 0 & 0 & 0 & 1 \end{pmatrix} \begin{pmatrix} 1 \\ 0 \\ 0 \\ 0 \end{pmatrix} = \begin{pmatrix} 5/3 \\ 4/3 \\ 0 \\ 0 \end{pmatrix}.$$

This, too, can be easily seen to satisfy the normalization condition.

3.7 The Lorentz Group*

One of the things you may have noticed about the Minkowski metric in 1+1 dimensional space is that, though it looks very similar to the Euclidean metric in two-dimensional ordinary space, boosts (with γ factors and whatnot) in the former don't look particularly similar to rotations (with trig functions) in the latter. In order to connect the two—and to revisit the idea of groups—let's compute the constraints on groups that keep the Minkowski metric invariant.

In terms of notation, consider a generic 2×2 transformation matrix,

$$\Lambda = \begin{pmatrix} a & b \\ c & d \end{pmatrix},$$

where all of the components are required to be real. The transformed metric can be computed as

$$\bar{\mathbf{g}} = \Lambda \mathbf{g} \Lambda^T$$

$$= \begin{pmatrix} a & b \\ c & d \end{pmatrix} \begin{pmatrix} 1 & 0 \\ 0 & -1 \end{pmatrix} \begin{pmatrix} a & c \\ b & d \end{pmatrix}$$

$$= \begin{pmatrix} a^2 - b^2 & ac - bd \\ ac - bd & c^2 - d^2 \end{pmatrix}$$

$$= \begin{pmatrix} 1 & 0 \\ 0 & -1 \end{pmatrix},$$

where I've implicitly insisted that the metric remain invariant under the transformation. This is satisfied if $a = d$, $b = c$, and

$$a^2 - b^2 = 1.$$

Given that these are the only two independent elements, this relation looks very much like the hyperbolic trig function:

$$\cosh^2 \theta - \sinh^2 \theta = 1.$$

Or, in other words, a boost can be expressed as

$$\Lambda^{\bar{\mu}}{}_{\mu} = \begin{pmatrix} \cosh \theta & \sinh \theta \\ \sinh \theta & \cosh \theta \end{pmatrix}. \tag{3.14}$$

In this form, a boost looks very similar to the rotation in 2D (eq. 1.9), except that, instead of trig functions, we've got hyperbolic trig functions.

This combination of hyperbolic trig function boosts in spacetime and trig function rotations in space is known as the **Lorentz group**, and it basically describes the symmetry of space and time. All of the physics that we've yet uncovered seem to be invariant under transformations in the Lorentz group: rotations (and displacements) in addition to boosts. The latter is, in fact, what Einstein's postulates were all about.

There's one final detail. It obviously doesn't matter much how you express it:

$$\cosh\theta = \gamma,$$

which, after some ugly algebra, inverts to

$$\theta = \ln\left(\sqrt{\frac{1+v}{1-v}}\right). \tag{3.15}$$

At small boosts, for $v \to 0$, we find $\theta \approx v$. However, as $v \to 1$, we get $\theta \to \infty$ (albeit much slower), allowing an arbitrary combination of boosts.

Looking Forward

Thus far, we've figured out how to transform space and time from one frame to another. But as we've already seen from the perspective of relativistic flying cubes, simply asserting length contraction and time dilation is insufficient for the task of accurately describing how things really *look* in a relativistic universe. In the next chapter, we'll explore this dilemma more fully. We'll talk about some classic puzzles and brainteasers in relativity, including a few paradoxes. And then we'll resolve those paradoxes.

3.8 Problems

1. A spaceship is traveling at a speed of $\frac{12}{13}$ c. If 1 hour passes on Earth, how much time passes on the spaceship's clock (as measured by people on Earth)?
2. To get you accustomed to computing gamma factors, compute γ for the following speeds:
   (a) $v = 300\,\text{m/s}$ (This is roughly the speed of sound on Earth. You'll need to do unit conversions, but you may find that your calculator needs to print a lot of digits.)
   (b) $v = 30\,\text{km/s}$ (This is the orbital speed of Earth around the Sun.)
   (c) $v = 0.01$

(d) $v = 0.5$

(e) $v = 0.9999$

3. Convert the following γ factors to speeds:

   (a) 1.001

   (b) 1.25

   (c) 5

   (d) 1000

4. You are inside a spaceship traveling at $v = 0.9$ and you throw a ball at $v = 0.5\, c$. What is the speed of the ball as seen from outside if it's thrown (a) forward or (b) backward?

5. Consider two events separated by $\Delta x = 5$ and $\Delta t = 4$ in the unprimed frame.

   (a) Compute the interval squared, Δs^2.

   (b) Compute $\Delta x^{\bar{\mu}}$ for the two events in a new frame such that the unprimed frame is moving $v = 0.8c$ to the right.

   (c) What is the interval in the new frame?

6. Consider a spaceship traveling to the right at $v = 0.9$.

   (a) Inside the spaceship, a toy spaceship flies to the right at $u_1 = 0.9$ with respect to the spaceship. How fast is the toy as seen from the outside of the spaceship?

   (b) The toy launches a tiny torpedo that moves at $u_2' = 0.9$ with respect to the toy. How fast is it seen from inside the spaceship? From the frame outside the spaceship? Note: I'm using primes because "unbarred" is the spaceship itself and "barred" is outside the spaceship, so I need a new, third frame.

   (c) The torpedo launches a laser $u_3'' = 1$ with respect to the torpedo. How fast is the laser as seen from the toy? the spaceship? outside the spaceship?

7. A spaceship travels at a speed of 0.5 to a Ross 154, a star 10 light-years away.

   (a) How long does the trip take according to the folks on Earth?

   (b) Draw the trip on a spacetime diagram (using the frame of reference of Earth).

   (c) Halfway through the trip, the pilot sends a signal (via light signal, naturally) back to Earth. Add that detail to your spacetime diagram.

   (d) Earth responds immediately. Does the pilot receive the response before reaching Ross 154?

   (e) What is the proper time of the trip?

   (f) Draw the trip (including the worldlines of Earth and Ross 154, and the communications between Earth and the spaceship) in the frame of reference of the ship.

8. Ignoring the time it takes to accelerate and decelerate, how fast would you need to travel to make it to the center of the galaxy (25,000 ly) in a proper time of

(a) 25,000 years? (Not useful unless you can upload your brain to a robot.)

(b) 10 years? What assumptions did you make in your calculation?

(Note: The latter answer especially will be relatively close to the speed of light, so don't just round up. You could express your answer as $v = 1 - x$, where x is some number.)

9. A spaceship is traveling at $v = 0.8$ in the $+x$-direction.

(a) What is the 4-velocity U^α of the ship?

(b) Calculate $\mathbf{U} \cdot \mathbf{U}$ directly (even if you know what it's supposed to be).

Note: I haven't used this notation a lot, but I want you to be familiar with it. The dot product of a 4-vector is just a shorthand:

$$\mathbf{U} \cdot \mathbf{U} = g_{\alpha\beta} U^\alpha U^\beta,$$

unlike 3-vectors, which have arrows above them.

(c) The spaceship accelerates to $v = 0.81$. What is the *new* 4-velocity, $U^{\alpha(new)}$?

(d) What is the *change* in 4-velocity, $\Delta U^\alpha = U^{\alpha(new)} - U^\alpha$?

(e) Calculate $\Delta U \cdot U$. As the step size, ΔU, gets smaller and smaller, how do you think this dot product would scale in terms of the step size?

10. You are docked at a space station. During battle exercises, two spaceships pass one another (and you) at high speeds. One, A, is traveling to the right at speed v (with respect to you), and the other, B, is traveling to the left at the same speed. You don't know the speed, but you *do* know that, according to the captains of the two ships, both ships are foreshortened by a factor of 2. If a standard spaceship length is supposed to be 100 m (that is, A and B both measure their counterparts as having 50 m spaceships), how long do *you* measure the spaceships to be?

FURTHER READINGS

- David Appell, "The Invisibility of Length Contraction," *Physics World* 32, no. 8 (August 2019): 41–45. A very nice, nontechnical discussion of the Penrose-Terrell effect, and how macroscopic three-dimensional objects would appear.
- Albert Einstein and Leopold Infeld, *The Evolution of Physics* (Simon and Schuster, 1938). The entire book, which tracks the development of modern physics, is worth a read, but students focusing on this chapter may want to look closely at Section III, "Field, Relativity," and in particular, the subsection "Time, Distance, and Relativity."
- Moses Fayngold, *Special Relativity and Motions Faster Than Light* (Wiley-VCH, 2008). Despite the provocative title, this is a very nice general introduction to

relativity, including both general discussions of Lorentz transforms and why the speed of light serves as an unbreakable barrier.

- H. A. Lorentz, *The Einstein Theory of Relativity* (Benediction Classics, 2012/1919). This is an interesting work, especially because a number of science historians give some measure of precedence (not to mention, credit) to Lorentz for the discovery of relativity. His insights and explanation and his explicitly giving Einstein credit for the theory are quite instructive.

Paradoxes

Former astronauts and twins Senator Mark Kelly (1964–) and Commander Scott Kelly (1964–). Scott spent over 500 days in space moving at high speeds and, through relativistic time dilation, aged several milliseconds less than his brother. Credit: NASA.

Now that we nominally understand how time and space *work* in special relativity, we need to reckon with some true oddities. This goes above and beyond the general insults to our intuition. It *feels* like time should run at the same rate for all observers, and yet in developing the Lorentz transforms, we see that is plainly not the case. And we'll find that relativity produces even more non-intuitive paradoxes. We'll spend this chapter developing and resolving those paradoxes.

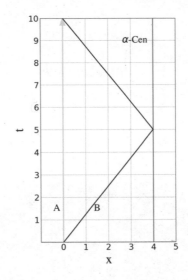

FIGURE 4.1. The spacetime trajectory of a twin traveling to α Centauri at 0.8 c, and returning at the same speed. As we saw in our discussion of proper time, this means that she ages *less* than the stay-at-home twin.

4.1 The Twin Paradox

We begin with one of the classics of relativity: The **twin paradox** [24]. In the simplest telling, we have the following situation. Two twins, Amy and Beth, are 20 years old. Beth gets on a spaceship and travels to α Centauri (4 light-years away), at a speed of 0.8 c (figure 4.1). By our (and Amy's) reckoning, the two-way trip takes

$$t_A = 2 \times \frac{4\,\text{yr}}{0.8} = 10\,\text{yr}.$$

By the time Beth returns, Amy is 30 years old.

On the other hand, Beth's clock runs slow by a factor of

$$\gamma = \frac{1}{\sqrt{1-v^2}} = \frac{5}{3}.$$

As such, she only ages

$$t_B = \frac{t_A}{\gamma} = 6\,\text{yrs}.$$

What's more, Beth only measures each leg of the trip to be 3 years. 3 years to travel 4 light-years! She *seems* to be traveling faster than the speed of light!

So far, so weird, but I haven't actually explained why it's a paradox. In our derivation of special relativistic time dilation, we made a big deal about two relativistic truths:

1. Einstein's second postulate suggests that inertial observers can't make measurements to distinguish whether they are the ones moving or standing still.
2. As a consequence, both Earth clocks and spaceship clocks should be seen to be running slow by observers in the other frame of reference.

At any given instant, we could imagine the situation from Beth's frame of reference, because as long as she's an inertial observer, it is equally valid to say that she is at rest and Amy (and the entire Earth) are in motion—and thus their clocks are running slow. But upon her return, we can say objectively that the traveling twin is younger than her sister, Amy.

How so?

Beth isn't an inertial observer. She needs to accelerate and then decelerate. We've violated the fundamental postulates of special relativity. But that doesn't mean that we need to throw out everything we learned about relativity in the process.

At any given moment, we can describe Beth's perspective in terms of a **momentarily comoving reference frame** (MCRF). That is, there's a frame that's moving at that exact velocity at that exact instant. An instant later, there's another MCRF. We can see the worldline of a continuously accelerated observer in figure 4.2.

We haven't encountered accelerations before, so you might be concerned about the units. In our natural units:

$$[a] = [T]^{-1}.$$

So for a spaceship, we might think about accelerations in terms of ly/yr^2 or yr^{-1}. Fortunately, the numbers turn out nicely. The most commonly used acceleration, g, turns out to be

$$g = 9.8\,m/s^2 \approx 1.03\,yr^{-1}. \tag{4.1}$$

Almost exactly 1! Another way of putting that is that you can get up to relativistic speeds in comfort (the back of your ship would feel like the floor, and inside, it would feel like Earth's normal gravity) in about a year.

Long paths in a spacetime diagram mean short proper times (figure 4.3). In this particular case, we imagine Beth accelerating at g up to $v = 0.8$, decelerating and reversing upon nearing α centauri, and then decelerating on Earth. Amy's worldline, meanwhile, is just a vertical line.

The faster the traveling twin accelerates up to coasting speed, the more the twin experiment resembles the idealized case in figure 4.1. We can play

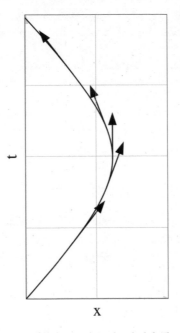

t

X

FIGURE 4.2. The worldline of a particle being accelerated to the left. The tangential arrows indicate the MCRFs at each instant.

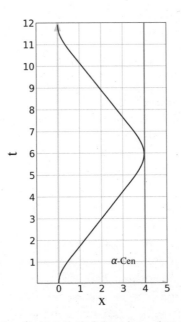

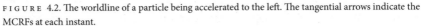

FIGURE 4.3. A "realistic" trip to α Centauri, in which the rocket accelerates to $v = 0.8$, coasts, decelerates and reverses direction upon reaching the destination, and then returns to Earth. The ages of the twins may then be compared directly.

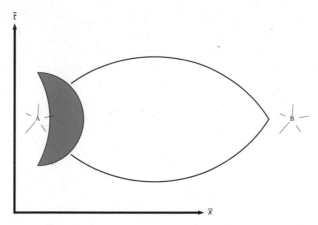

FIGURE 4.4. A slightly updated version of the train paradox [12], in which two explosions are seen outside a moving spaceship. In the reference frame of the space station, the two explosions are simultaneous.

all sorts of games with these numbers. We could make the trip longer, for instance, 50 or 100 light-years. We could speed up the traveling twin to 0.9 or 0.99 or any number (less than 1) that we like.

For instance, a trip to a destination 100 ly away at $v = 0.99$ will take about 202 years according to Earth time, but only 28.5 years on the spaceship. Indeed, this could be the whole *point* of the trip. You could travel two centuries into the future in a human lifetime.

There is a downside, however. While relativistic speeds are an effective method for time travel into the future, there's no way to get back to the time you started from.

4.2   The Train Paradox

In the previous paradox, we concerned ourselves with the flow of time with regard to acceleration, but even in the limit of simple inertial frames, comparisons of time present a problem. Einstein and others fretted about the physics of **simultaneity**. He posed this riddle [12]:

> We suppose a very long train traveling along the rails with the constant velocity
> v.... Are two events (e.g., the two strokes of lightning A and B [or the front
> and back of the train]) which are simultaneous with reference to the railway
> embankment also simultaneous relatively to the train? We shall show directly
> that the answer must be in the negative.

As is our custom, we'll consider the same basic riddle, but set in a spaceship (figure 4.4).

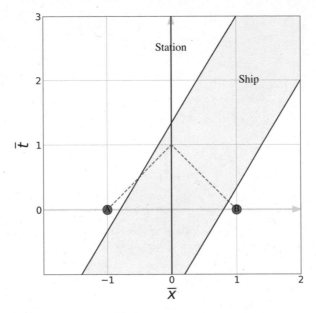

FIGURE 4.5. The train paradox spacetime diagram as seen from the space station (barred) reference frame. Two simultaneous explosions are detected by an observer standing at the origin. Meanwhile (as drawn), the spaceship flies by at $v = 0.6$.

As the spaceship flies past a space station, the station inhabitants see two simultaneous explosions (labeled A and B) at the front and back of the ship. Einstein wanted to know whether the explosions would also appear simultaneous to the spaceship. The easiest way to envision this is to imagine an observer situated at the midpoint of the ship at the instant that the explosions go off. We'll call that the origin (figure 4.5).

The explosions are simultaneous as seen from the station, but what about the perspective of the ship? We can answer this question easily enough using the inverse Lorentz transforms that we developed in Chapter 3 (eq. 3.9). Using $\Delta \bar{x} = L$ and $\Delta \bar{t} = 0$ (the definition of simultaneity), we find

$$\Delta x^\mu = \Lambda^\mu_{\ \bar{\mu}} \Delta x^{\bar{\mu}}$$

$$= \begin{pmatrix} \gamma & -v\gamma \\ -v\gamma & \gamma \end{pmatrix} \begin{pmatrix} 0 \\ L \end{pmatrix}$$

$$= \begin{pmatrix} -v\gamma L \\ \gamma L \end{pmatrix}.$$

The time interval as seen within the ship is no longer zero! Indeed, from the perspective of the ship, explosion B occurs *before* explosion A. We can see what it looks like from the spaceship's perspective in figure 4.6.

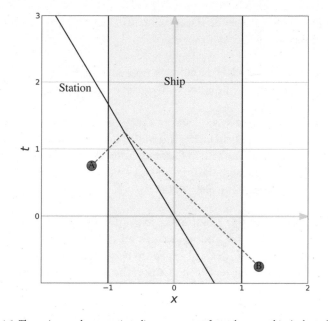

FIGURE 4.6. The train paradox spacetime diagram as seen from the spaceship (unbarred) reference frame. The "simultaneous" events are not seen as such, according to the spaceship. They're not even centered.

But wait. There's more! A ship traveling in the opposite direction would see the events occur in the opposite order (A then B). This means that the time order of events is just a matter of perspective. One might even say that it's all relative. That does not mean that it's completely arbitrary. While the order of the explosions is up to interpretation, it doesn't actually matter which one comes first. The two events are spacelike separated (as drawn, $\Delta s^2 = -4$), which means that no matter how you slice it, neither event can affect the other, because the signal from one to the other can never exceed the speed of light.

4.3 The Ladder Paradox

We've focused our discussion on paradoxes arising from time dilation thus far, but length contraction isn't exempt from messing with our physical intuition. In 1961, Wolfgang Rindler [37] began a paper with a silly story:

> A certain man walks very fast—so fast that the relativistic length contraction makes him very thin. In the street he has to pass over a grid.* A man standing at the grid fully expects the fast thin man to fall into the grid. Yet to the fast man

_____
*A grate, in the United States

the grid is much narrower even than to the stationary man, and he certainly does not expect to fall in. Which is correct?

Rindler described a class of paradoxes now known as the ladder paradox (sometimes called the barn paradox) because of similar setups:

> Imagine a ladder of fixed length L and a garage of just a tiny bit smaller depth so that the ladder can't quite fit inside. In addition to the garage door, the garage has a back door. If you run toward the garage with the ladder at a high enough speed, could you close both the front and back doors of the garage so that, for an instant, the ladder is completely inside?

A simplified version of this situation appears in figure 4.7.

The mechanics are straightforward. To an observer in the garage frame (the barred frame), the ladder will appear to be of length L/γ, as is shown in a spacetime diagram in figure 4.8. In terms of the diagram, the series of events is as follows:

1. Both doors are open, and the front of the ladder enters the garage around $\bar{t} \approx 0.5$.
2. At $\bar{t} = 1.25$, the back of the ladder goes through the front door.
3. A moment later, both doors shut.

What happens next is outside the scope of the problem, but it's clear that if you have a ladder moving at $v = 0.8\,c$, (even if it did fit inside the garage), as a practical matter, it's going to bust through the back door a fraction of a second later.

Indeed, when Rindler wrote the original problem, he was focused on *exactly* this sort of implication. In relating his description of the man walking

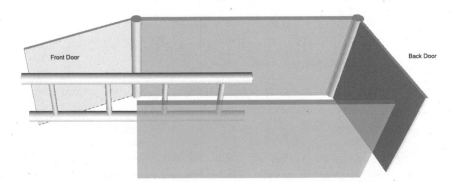

FIGURE 4.7. The ladder paradox. A ladder is brought into a garage at relativistic speeds. If (at rest) the ladder and garage are of the same length (or even if the ladder is slightly longer) is it possible to use length contraction to close both the front and back door at the same time?

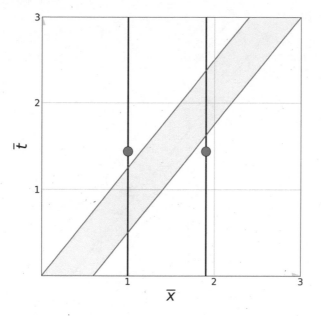

FIGURE 4.8. From the perspective of the garage, which has a depth of 0.9. The incoming ladder has a rest length of $L = 1$, and a gamma factor of $5/3$. Is there a moment when both the front and back doors can be simultaneously closed? There sure is! The dots indicate just such points.

over the grate, I omitted the final sentence in his abstract: "The answer hinges on the relativity of rigidity." He focused on the question of the front of the man falling over the edge of the open grate while the back was still firmly over the grate, and whether you can really bend something that rigid.

The central *paradox* concerns the fact that, from the ladder's perspective, the garage appears length contracted, and there should be no way that the ladder fits inside entirely. How do you resolve the issue of the garage frame thinking that the ladder is foreshortened, but the ladder frame sees the garage as foreshortened?

If the ladder is *really* smaller than the garage, then wouldn't that mean that there's a preferred frame of reference and you can say with absolute certainty that one is moving and one is sitting still? Einstein's postulates (and all of the work we've done so far) demand that there is no measurement that can determine the absolute state of motion of either ladder or garage.

We might ask the question of what happens when the ladder hits the back wall. Even if I insist that the back door is perfectly rigid, there's still the question how the ladder "knows" to stop. The front of the ladder hits the wall, but the forces propagate backward through the ladder and only hit the back of the ladder after a time $\Delta t > L_0$ in the ladder reference frame. It can't

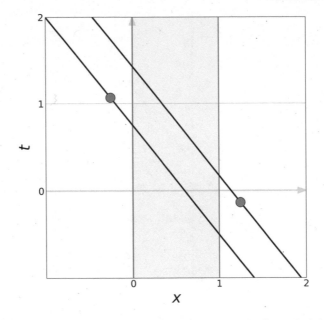

FIGURE 4.9. The ladder paradox from the ladder's perspective. In that frame, the ladder is never completely inside the fully closed garage.

be instantaneous. Indeed, it's not (figure 4.9). The sequence of events is as follows:

1. The back door is shut around the same time the front of the ladder ($x = 1$) is passed by the front of the garage.
2. The back of the ladder collides with the closed back door (or vice versa).
3. The front door is shut.

At no time (from the ladder's perspective) is the ladder completely inside the garage.

There is a rich history of similar paradoxes in special relativity in which the "story," such as it is, needs to be told from several perspectives. One particularly interesting example is the flickering light bulb paradox [41]. A light bulb is set up on a simple broken circuit. Meanwhile, a bridge, closing the circuit, moves to the right at relativistic speeds, as shown in figure 4.10.

The connector (in its own rest frame, naturally) is of some length L, as is a "gap" in the circuit. In the connector frame, the gap appears shortened and, thus, we might assume that the circuit is always closed. The light stays lit without interruption. In the circuit frame, the connector appears shortened, and for some period of time, the circuit is open, and the light flickers off and

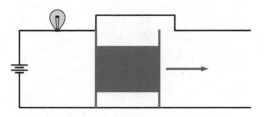

FIGURE 4.10. The flickering light bulb paradox. Can a connector moving at relativistic speeds close a circuit when it appears length contracted in the frame of the circuit itself?

then on again. Which is it? You'll have the opportunity to explore this paradox in the end-of-chapter problems.

4.4 Ehrenfest Paradox

Shortly after Einstein proposed special relativity, Paul Ehrenfest wrote a brief letter [5] pointing out a possible paradox.* Ehrenfest invites us to imagine a large, rigid (there's that word again!) spinning disk (figure 4.11). If the disk rotates at an angular velocity of ω, a point at a distance from the center, r, will have a speed of

$$v = r\omega.$$

There is one obvious problem if the disk gets large enough. If $r > \frac{1}{\omega}$, the edge of the disk will apparently move at $v > 1$. That clearly isn't allowed. At a minimum, because the electromagnetic forces holding the disk can't travel that fast, the assumptions of rigidity would break down.

But even at subluminal speed, there seems to be a problem. If the rotator (disk) really is rigid, then before it starts spinning up, the outer circumference is $2\pi R$.

Once the disk starts spinning up, there will be a centrifugal force pushing outward. If anything, you'd expect the edge to get stretched from ordinary Newtonian mechanics, but as we're insisting that the rotator is rigid, the edge neither expands nor contracts.

But here's where relativity comes into it. To an observer standing nearby, the edge appears to be Lorentz contracted along the direction of motion by a factor of

$$C = \frac{C_0}{\gamma} = C_0 \sqrt{1 - r^2\omega^2}.$$

*It's also interesting to note that Ehrenfest didn't acknowledge Einstein at all, but rather attributed the ideas as "Minkowskian."

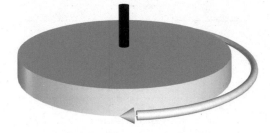

FIGURE 4.11. A large, rigid rotator (disk), spinning at a fixed angular velocity, ω.

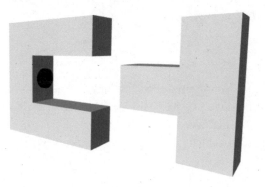

FIGURE 4.12. The lock and key paradox in which a lock and key approach each other at relativistic speeds, resulting in the lock appearing shallower from the key perspective. Can the key unlock the lock?

The circumference should somehow be smaller than the number that you get from Euclidean geometry. The other interpretation is that there is no such thing as a strict rigid rotator, because even at nonrelativistic speeds there's going to be *some* contraction of the outer rim. As we will see in Chapter 8, this paradox is precisely the sort of argument that Einstein used to motivate his development of general relativity.

The resolution to this paradox centers on the idea of rigidity. A rigid rotator requires the relative positions of all points to remain fixed with respect to one another, but what Ehrenfest was really showing was that rigidity has no real place in special relativity. The outer edge would be drawn out by centripetal force even though, in the local frame, it would appear Lorentz contracted.

There are a lot of paradoxes in special relativity that ultimately rest on the question of whether it *really* makes sense to think of anything as rigid when moving at relativistic speeds. Take the **lock and key paradox**, [35] which imagines a key moving toward a lock with the key nominally too short to hit the back of the lock mechanism (figure 4.12).

At first blush, the lock and key paradox seems similar to the ladder paradox. But unlike the ladder paradox, it revolves around the question of whether the "front" of the key is stopped when the back of the key collides with the lock. In a perfectly rigid system, it would. However, as with the Ehrenfest paradox itself, it centers around the fact that bodies simply cannot respond instantaneously to changes.

Likewise, J. S. Bell [3] popularized a paradox involving two accelerating ships with a rigid tether between them, focusing on the question of whether the tether would snap provided the ships' acceleration were perfectly synchronized. His corresponding discussion of this and other related paradoxes is sufficiently approachable that I've included it in the readings below.

4.5   Superluminal Motion

One of the biggest downers in this book is that we've definitively shown that you can't move faster than light, somewhat limiting prospects for interstellar travel. Even though you can make it to other stars in a human lifetime, doing so requires time dilation factors so large that you will essentially have pushed yourself into the future with no way to return to your loved ones in the past.

But there are *apparent* ways to produce faster than light speeds, and we saw a hint of this in Section 3.5 in our discussion of the Penrose-Terrell effect. In relativistic jets emanating from quasars (and a few other high-energy astrophysical systems), we sometimes see an apparent **superluminal motion**—particles traveling faster than the speed of light. This effect has been observed for more than a century, even predating relativity itself by a few years (as in 1901, when Jacobus Kapteyn observed the detritus left by the exploding nova GK Persei [23]), and has been seen in myriad systems since.

To understand both the apparent paradox and its resolution, we need to say a few words about how we measure the motions of stars or other astronomical sources. As we will discuss quantitatively in the next chapter, the redshift of a moving source allows you to measure the component of velocity, v_r, along the line of sight.

We can also measure the **tangential velocity** (perpendicular to the line of sight) by watching an object move across the sky. For an object at distance d,

$$v_t = \mu d,$$

where μ is the so-called **proper motion**—the measure of angle (in radians to make everything work neatly) per unit time. Naturally,

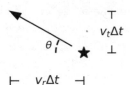

FIGURE 4.13. The motion of a "superluminal" cloud of gas starting at the star. The velocity is at angle θ with respect to the line of sight.

$$v^2 = v_r^2 + v_t^2,$$

as usual.

Suppose that this source moves with a large velocity, principally toward Earth. This might be (as the first observed objects were) jets coming out of black hole accretion disks (figure 4.13).

After some time, Δt, the object will have moved in a tangential direction:

$$\Delta x = v_T \Delta t.$$

In the radial,

$$\Delta r = -v_r \Delta t.$$

But wait! Since the endpoint of the trajectory is closer to us than it originated, it'll reach us earlier than expected:

$$\delta t = \frac{\Delta r}{c} = -\frac{v_r}{c} \Delta t.$$

To us here on Earth, the "measured" transverse velocity is thus

$$\tilde{v}_t = \frac{\Delta x}{\Delta t_{apparent}}$$
$$= \frac{v_t \Delta t}{\Delta t - v_r \Delta t}$$
$$= \frac{v_t}{1 - v_r}$$
$$= \frac{v \sin \theta}{1 - v \cos \theta}.$$

By inspection, some combination of v and θ, the cloud could produce apparent speeds faster than 1. How fast? Well, for fixed v, the value of maximum

θ can be computed by taking the derivative and setting it to zero. Some straightforward, but tedious, calculus and algebra produce:

$$\cos \theta_{max} = v.$$

Likewise,

$$\sin \theta = \sqrt{1 - \cos^2 \theta} = \sqrt{1 - v^2}.$$

Plugging this in, we get an interesting relation:

$$\tilde{v}_t = \frac{v \sin \theta}{1 - v \cos \theta}$$

$$= v\gamma.$$

What a tidy result!

Note that for $v \ll 1$, the maximum speed is for $\theta = \pi/2$. Obviously, that corresponds to an object moving perpendicular to the line of sight. Naturally, its proper motion will be greatest. For $v \to 1$, the max is for $\theta \to 0$, which makes sense because a beamed object would appear to have a maximum speed.

All of this is to say that, while relativity doesn't actually create the paradox of superluminal apparent velocities, it does shed some light on it—so to speak.

Looking Forward

The focus of this chapter and those preceding was to get you comfortable with the sorts of paradoxes that arise from the transformation of one frame to another. This *weirdness* normally takes center stage in a discussion of relativity, but in some sense, our broader goal is to simply get our bearings. As we move forward, we're going to rely on an understanding of different perspectives to provide the foundation for energy and momentum. We'll focus less on transforming between frames (which will be a relief if you've grown tired of drawing spacetime diagrams) and rather will use the fact that we *could* transform between frames if we wanted to—and from that, we're going to find some very interesting results.

4.6 Problems

1. The center of the galaxy is roughly 20,000 light-years away. Roughly how fast would you need to travel to complete a round-trip journey in a human lifetime (assume 80 years)? How much time would have passed on Earth? Ignore the time required for accelerate.

2. As a first step in thinking about how acceleration affects time dilation, consider an astronaut on a rocketship accelerating up to decidedly sub-light speeds under constant acceleration:

$$v = at.$$

Noting that

$$d\tau = \frac{dt}{\gamma}$$

and using the relation $(1 + x)^n \approx 1 + nx$ for small values of x, integrate explicitly to find a relationship for the fractional deficit in proper time measured on the spaceship compared to that passing back on Earth.

3. A UFO flies past Earth at $t = \bar{t} = 0$, and then flies past the Moon (approximately 1 light-second from Earth according to earthlings) at a speed of $v = 0.6$. The Earth-Moon frame is unbarred, with $x = 0$ corresponding to Earth.

   Both earthlings and moon dwellers want to contact the alien, so at time t_1, both simultaneously send a hail to the UFO.

   (a) At what time t_1 do the earthlings and Moon people send their signal so that they arrive simultaneously?

   (b) Sketch the worldlines of the Earth, Moon, UFO, and hails on a spacetime diagram.

   (c) Even though the humans believe both signals to be sent simultaneously, the aliens don't agree. Which signal do they believe was sent first, and by what margin?

   (d) Draw the same events from part (b) as seen on the UFO.

4. Consider the flickering light bulb paradox (figure 4.10). Both the gap and the connector have the same width (each in their own rest frame) L. The connector moves at speed v. Throughout, $x^{\bar{\mu}}$ corresponds to the connector's coordinates and x^{μ} to the bulb's. The height of the circuit is small enough that the vertical electrical travel time may be assumed to be zero.

   (a) At time $t = \bar{t} = 0$, the left-hand side of the connector aligns exactly with the left-hand side of the gap. At what time (in the unprimed frame) does the right-hand side of the connector make contact with the right-hand side of the gap? Note that the spatial coordinate for this event is necessarily $x = L$.

   (b) Assuming that current travels at a speed $v_c = 1$ (not a realistic assumption, but it doesn't actually matter), how long is the "flicker" as seen in the rest frame of the light bulb?

   (c) What are the spacetime coordinates, $x^{\bar{\mu}}$, for the connector making contact with the right-hand side of the gap (e.g., the event in part [a])?

   (d) How long does the flicker last in the frame of the connector? (Relevant to whether this is a paradox, is this a positive number?)

5. The discussion of the Ehrenfest paradox is intended to lead into the the development of general relativity in Chapter 8. In particular, consider a rigidly

rotating disk with angular velocity ω in the limit that the tangential velocity $\omega r \ll 1$.

(a) Compute an approximate Lorentz factor, γ, as a function of r. You should use the same Taylor expansion as in Problem 1.

(b) Calculate the centrifugal acceleration as a function of r, as experienced by an observer on the disk, and relate it to the apparent potential $a_r = -\frac{d\Phi}{dr}$. Calculate Φ. This, too, should be a function of only r.

(c) Express the γ factor as a function of Φ.

FURTHER READINGS

- *American Journal of Physics*. While most of my recommendations are for individual books and articles, the *American Journal of Physics* is the journal for the American Association of Physics Teachers, and thus has a great many articles specifically focused on teaching key physics concepts. There are a number of great articles related to classic paradoxes, including the following:
    - Roberto de A. Martins, "Length Paradox in Relativity," *American Journal of Physics* 46, no. 6 (1978): 667–70.
    - Robert H. Romer, "Twin Paradox in Special Relativity," *American Journal of Physics* 27, no. 3 (1959): 131–35.
    - Robert Perrin, "Twin Paradox: A Complete Treatment from the Point of View of Each Twin," *American Journal of Physics* 47, no. 4 (1979): 317–19.
    - Evan Pierce, "The Lock and Key Paradox and the Limits of Rigidity in Special Relativity," *American Journal of Physics* 75, no. 7 (2007): 610–14.
- J. S. Bell, "How to Teach Special Relativity," *Progress in Scientific Culture* 1, no. 2 (1976): 1–13. PDF reprints are readily available. The author is most well known as the originator of Bell's inequality in quantum mechanics, which showed how one might distinguish between classical and quantum systems. Bell's work is very readable and, in this case, addresses dynamical paradoxes in physical systems—a very useful lead-in to discussions of relativistic momentum.
- Edwin F. Taylor and John Archibald Wheeler, *Spacetime Physics* (W. H. Freeman, 1992). I have noted this book for further reading earlier in this discussion, but I would especially point the reader to the discussion of the flickering light bulb on pp. 186–87. A particular strength of the text is an almost encyclopedic inclusion of classic paradoxes.

5

Momentum and Energy

Marie Skłodowska-Curie (1867–1934) circa 1907. Curie isolated polonium, but more generally, she is credited with pioneering work in the understanding of radioactivity, one of the most direct (and, sadly, in the case of Curie, deadly) manifestations of the conversion of mass to energy in terrestrial conditions.

5.1 Doppler Shifts

Light is weird. In the seventeenth century, Christiaan Huygens [22] showed incontrovertibly that light has the properties of a wave: it exhibits interference and diffraction and has a well-defined wavelength.

Then came 1905, Einstein's annus mirabilis, his miracle year. Einstein wrote his foundational papers on relativity [7, 9], proved the existence of atoms [10], and, most important for the present discussion, showed just as

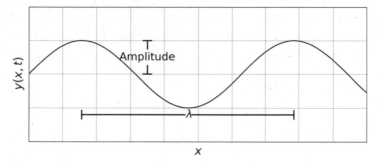

FIGURE 5.1. The wavelength, λ, and amplitude of a propagating wave.

incontrovertibly that light must be a particle and not a wave [8].* The photo-electric effect involves shining high-energy light onto metals and knocking out electrons in the process. Einstein's contribution was showing that this could only be the case if light came in the form of discrete packets—photons. This work was also cited in his 1921 Nobel Prize "for his services to theoretical physics, and especially for his discovery of the law of the photoelectric effect."

This isn't a book about quantum mechanics, so we can content ourselves with the knowledge (if not the deep understanding) that light has properties of both a particle and a wave, and, at the very least, can try to figure out how to relate the properties of the two within the context of relativity.

A simple, sinusoidal propagating wave (figure 5.1) can characterized by three numbers:

1. Frequency, ν
2. Wavelength, λ
3. Propagation speed, $v = c^\dagger$

These three numbers are related via

$$c = \lambda\nu.$$

A wave also has amplitude—how high the wave is. In the case of an electro-magnetic (light) wave, amplitude corresponds to the strength of the electric field. In the case of almost all waves, the strength of the electric field is

---

*Quantum mechanics, which is sadly outside the scope of this book, "resolves" the conflict by showing that light is both wave and particle.

†In general, there are actually two "speeds" for wave propagation, a group velocity and a phase velocity, but the distinction isn't important in this context and is nonexistent for light moving through a vacuum

proportional to the square of the amplitude: with electromagnetic fields, $E^2 + B^2$ gives the energy density of the field; with quantum mechanical waves, $|\psi|^2$ gives the probability of finding a particle in a particular place.

As paradoxical as it seems, photons have both the properties of particles (discrete momentum, countability) and waves (interference, wavelength), and the two can be related via the handy equation

$$E = h\nu.$$

For instance, red light has a wavelength of $\lambda \approx 700\,\text{nm}$, so

$$\nu_{red} = \frac{c}{\lambda} = 4.3 \times 10^{14}\,\text{Hz},$$

with longer (redder) wavelengths corresponding to lower frequencies, and shorter (bluer) wavelengths corresponding to higher frequencies.

To visualize the propagation of electromagnetic waves in a (now) familiar context, let's plot the successive crests of a wave (figure 5.2). To an observer standing at rest at the origin, a peak is emitted every $\Delta t = 1/\nu$ and then proceeds rightward at a 45-degree angle in the diagram. The spatial separation of the peaks at fixed time is a measure of the wavelength. A distant but relatively stationary observer then absorbs the light, measuring an identical frequency at the detector.

We developed the Lorentz transforms to convert from one frame to another. So we'll do precisely that, and ask what happens if we take the wave from figure 5.2 and observe while fleeing from the source at speed, v. We add a moving observer (drawn at $v = 0.5$) to the right-hand panel of figure 5.2.

The observer intercepts the first crest at (x_1, t_1) and the second crest at (x_2, t_2). Because the observer is moving (and the coordinates are expressed in the unprimed frame),

$$x_2 = x_1 - \frac{1}{\nu} + (t - t_1),$$

where the frequency and wavelength are related in natural units via $\lambda\nu = 1$.

But as the observer themselves is traveling at speed v, we also know

$$x_2 = x_1 + v(t - t_1).$$

Setting the two previous equations together, we get

$$\Delta x^\mu = \frac{1}{\nu}\begin{pmatrix} \frac{1}{1-v} \\ v\frac{1}{1-v} \end{pmatrix}.$$

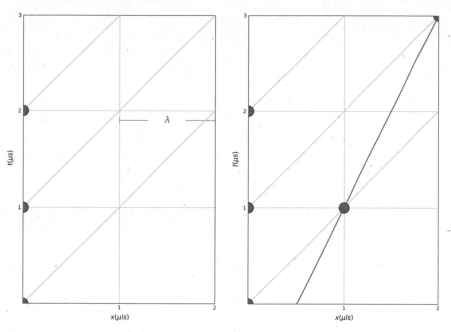

FIGURE 5.2. Left: A light wave of $\lambda = 1$, $\nu = 1$ is emitted from the origin. Peaks are emitted at $t = 0, 1, 2 \ldots$ propagating rightward at c. At a fixed time, the distance between the peaks can be measured ($\lambda = 1$). A distant observer then measures the period as $P = 1$, or $f = 1/P = 1$. Right: The moving observer intercepts the electromagnetic crests at different times and positions. All units are in the emitter frame.

This isn't how the observer sees things. The proper time between pulses will be

$$\Delta \tau^2 = \frac{1}{\nu^2} \left(\frac{1}{(1-\nu)^2} - \frac{\nu^2}{(1-\nu)^2} \right)$$

$$= \frac{1}{\nu^2} \left(\frac{1+\nu}{1-\nu} \right),$$

or equivalently,

$$\nu_{obs} = \nu \sqrt{\frac{1-\nu}{1+\nu}} \tag{5.1}$$

and

$$\lambda_{obs} = \lambda \sqrt{\frac{1+\nu}{1-\nu}}. \tag{5.2}$$

This is known as a **Doppler shift**, or, in this particular case, a **redshift**. This shift can be seen most simply via the dimensionless quantity z (often referred to as *the redshift*):

$$z \equiv \frac{\lambda_{obs} - \lambda}{\lambda} = \sqrt{\frac{1+v}{1-v}} - 1. \tag{5.3}$$

A positive (receding) relative velocity results in $z > 0$, which is equivalent to saying that an observer moving away from the source sees a longer wavelength. As with everything in relativity, it does not actually matter whether it's the source or the observer doing the moving. The velocity in this context represents only the *relative* motion. Likewise, for negative velocities, the observed wavelength is **blueshifted.** The ability to work out the speed of moving objects based on emitted or reflected frequencies of light is precisely the mechanism that allows you to be caught via radar in a speed trap on the highway.

For one final application, let's consider the case where $v \ll 1$. In this situation, we apply the handy Taylor series (eq. 3.3) to equation 5.3. If v is small, we can show

$$z \approx v.$$

In this limit, the redshift is just the fraction of the speed of light that the source and receiver are moving away from one another.

Example: Speed Traps

Police often use radar guns to measure the speed of oncoming cars. Suppose a car is traveling at 90 mph toward the speed trap. The speed gun emits light with a frequency of 10.525 GHz. What is the *difference* between the measured frequency of the oncoming car and the reference frequency?

Solution
While the numbers were given in mph, the speed is better written as 40.2 m/s, or even better yet, 1.34×10^{-7} in natural units, which is patently nonrelativistic. Thus, $v = z$ and

$$\Delta \nu = z \nu_0$$
$$= (1.34 \times 10^{-7})(1.0525 \times 10^{10}\, \text{Hz})$$
$$= 1411\, \text{Hz}.$$

5.2 Einstein's and Newton's Laws

Even though this is a physics text, we've really been doing a lot of math—math that admittedly was strongly motivated by a couple of basic ideas in physics. In particular, we (following Einstein) have taken to heart Galilean relativity and Newton's first law [31], though the two say much the same thing:

> Law I. Every body perseveres in its state of rest, or of uniform motion in a right line, unless it is compelled to change that state by forces impressed thereon.

Or, as it's normally phrased in introductory physics classes, "Objects in motion stay in motion unless acted upon by another force." This led, as we saw in Section 2.4, to Einstein's first postulate of special relativity, which demanded that the laws of physics are the same in all inertial reference frames. His addition of the second postulate—the constant speed of light—was what ultimately introduced γ and forced us to abandon a fixed flow of time.

From the rules governing tensors and our choice of the Minkowski metric, Einstein's postulates and Newton's first law may be combined into a single idea:

> All the laws of physics should be written in a Lorentz-invariant way.

What that means is that, as long as all of the equations are written using the tensor rules we've discussed, they're guaranteed to satisfy Einstein's postulates. It also means that the nonrelativistic version of Newton's second law,

$$F^i = m\frac{d^2x^i}{dt^2},$$

can't be correct for a variety of reasons. For one, it can't be the most fundamental relation, since the index is only over the spatial dimension, but we're really looking for a Greek index (space and time). For another, the derivatives are calculated with respect to t, and t is a frame-dependent variable. Rather, a we might expect relation of the form

$$f^\mu = \frac{dp^\mu}{d\tau},$$

where f^μ is a forcelike 4-vector, with the $\mu = 0$ timelike term acting like a power. As we will see in Chapter 7, this is precisely the form of the Lorentz force law of electromagnetism.

Regardless, Newton's third law of motion remains frame invariant in relativity. It reads:

Law III. To every action there is always opposed an equal reaction: or, the mutual actions of two bodies upon each other are always equal, and directed to contrary parts.

This corresponds to conservation of linear momentum, where the nonrelativistic version of the momentum is

$$\vec{p}_{NR} = m\vec{v}.$$

Whatever else we derive, the low-speed limit of momentum must ultimately yield this same result, just as the limiting case of the Lorentz transforms were the same as the Galilean transforms.

Consider two equal mass particles, initially at rest and bound to one another with massless cords. Left to their own devices, the two particles would repel one another (figure 5.3).

As everything is at rest, the total momentum of the system is necessarily zero. The connecting cord is cut, and the two particles fly off in opposite directions at speed v (figure 5.4). Symmetry and Newton's third law demand that they be equal and opposite. For our purpose, we'll assume that the initial mass of the combined system is $M = 2m$. As we will see, this is a bad assumption.

The physics in the rest frame of the original apparatus are pretty straightforward, but let's consider what they would look like to an observer traveling to the left with velocity $-v$. In that case, the initial apparatus will have $\bar{u} = v$ (the entire contraption is moving to the right). In that frame, the apparatus has a nonrelativistic momentum:

$$\overline{P}_I = Mv = 2mv. \tag{5.4}$$

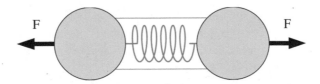

FIGURE 5.3. Two equal masses, m, are repulsive but held together by a massless cord. The system is at rest, so its momentum is zero.

FIGURE 5.4. The outgoing particles have equal and opposite velocity.

E_0 E_0

FIGURE 5.5. A photon of mass M decays into two photons of energy E_0.

In the barred frame, the first particle is stationary (by design), but the second particle has an apparent velocity of

$$\overline{u}_2 = \frac{2v}{1+v^2},$$

using our velocity addition rule. Thus, if the nonrelativistic momentum were correct, we'd have

$$\overline{p}_2 = \frac{2mv}{1+v^2} \neq \overline{P}_I = 2mv.$$

In the boosted frame, momentum doesn't seem to be conserved. We've made a mistake. Actually, we've made a few.

First, $p \neq mv$. And second, the mass of a composite particle is not just the sum of its masses.

5.3 Mass Energy

5.3.1 FROM MASS TO LIGHT

The time has finally come to derive the most famous equation in physics. You know the one I'm talking about.

From our previous discussion, we've already found that mass isn't necessarily conserved in relativity, but we don't actually know the exchange rate between mass and energy. To figure that out, let's consider a particle that decays into two photons (figure 5.5).

We don't know anything (yet) about photon momenta, but since the two travel in equal and opposite directions, their net momentum must be zero. That's perfect, because the initial particle is at rest as well. We're now going to make a fairly big leap—one that will turn out to be correct. It may be that it's possible to convert the mass of particles into energy. We'll call this **rest energy**, and even though we don't (officially) know what it is, it can be related to the photon energy:

$$2E_0 = E_M.$$

That's about all we can get from this frame of reference. Now, let's boost the frame to the right at some nonrelativistic speed v (figure 5.6).

E_L E_R

FIGURE 5.6. The same events shown in figure 5.5 but in a boosted frame. Note that the right-going photon appears blueshifted (higher frequency), while the left-going photon appears redshifted.

Provided the velocity is nonrelativistic, we can exploit our knowledge of classical mechanics to figure out the energy of the particle in the moving frame:

$$E_{particle} = E_M + \frac{1}{2}Mv^2.$$

The photons are a little more complicated. Einstein showed that the energy of a photon is related to its frequency:

$$E = h\nu.$$

But, as we've also seen, a blueshifted photon has a frequency shift of

$$\nu_R = \nu\sqrt{\frac{1+v}{1-v}}.$$

Multiply left and right by a factor of h, and, for the right-going (blueshifted) photon, we get

$$E_R = E_0\sqrt{\frac{1+v}{1-v}}.$$

For the left-going (redshifted) photon, we get

$$E_L = E_0\sqrt{\frac{1-v}{1+v}}.$$

The total energy of the photons in the boosted frame is

$$E = E_0\sqrt{\frac{1+v}{1-v}} + E_0\sqrt{\frac{1-v}{1+v}}$$

$$= E_0\sqrt{\frac{(1+v)^2}{(1-v)(1+v)}} + E_0\sqrt{\frac{(1-v)^2}{(1+v)(1-v)}}$$

$$= \frac{2E_0}{\sqrt{1-v^2}}$$

$$= E_M\gamma.$$

Our gamma factor strikes again! And note that this part of the derivation is exact.

But there's more! We can use the nonrelativistic limit to figure out what E_M actually is. First, we use our favorite Taylor expansion (eq. 3.3) to estimate the nonrelativistic gamma factor (eq. 3.4):

$$\gamma \approx 1 + \frac{1}{2}v^2,$$

so

$$E_M\gamma \approx E_M + \frac{1}{2}E_M v^2.$$

But we already saw that the energy of the particle in the boosted frame was $E_M + \frac{1}{2}Mv^2$, which gives us (drumroll, please):

$$\boxed{E_M = M}. \tag{5.5}$$

Putting in the factors of c to make it more familiar:

$$E_M = Mc^2.$$

We have even more than that. Because our calculation of the photon energy in the boosted frame was exact, we have a general relationship for the total energy of a moving, massive particle:

$$E = M\gamma = Mc^2\gamma. \tag{5.6}$$

Since the rest energy is M, this gives a general form of the kinetic energy as

$$E_K = M(\gamma - 1), \tag{5.7}$$

which, as we've seen, gives us the familiar nonrelativistic kinetic energy in the low speed limit.

5.3.2 IMPLICATION: MASS IS ENERGY

That was an awful lot of algebra, but it produced a remarkable result: the creation or annihilation of mass produces a huge amount of energy.

We've calculated mass-energy equivalence before, when we were first introduced to natural units. But now they're taking on a whole new meaning. A 70 kg (154 pound) person, for instance, has a mass-energy of

$$E = (70 \, \text{kg})(3 \times 10^8 \, \text{m/s})^2 = 6.3 \times 10^{18} \, \text{J},$$

which is a big enough quantity that it's really hard to picture. It's about 30 times more than the most powerful nuclear bombs ever created, but it's also a little bit confusing, because there's really no danger of all of your matter being converted into energy any time soon.

A somewhat more realistic example comes from looking at the annihilation of matter and antimatter. For instance, if an electron collides with a positron, we get

$$e^- + e^+ \to 2\gamma,$$

where γ in this context means photons. How energetic are those photons? For an electron,

$$m_e c^2 = 511 \text{ keV}.$$

For context, a photon of visible light has an energy of about 2 eV, which means that an electron contains about a quarter of a million times more energy.

This also works the other way around. If you have a sufficiently energetic pool of photons (that is, in a region of space or time that is sufficiently hot), you can create a particle and antiparticle pair:

$$\gamma + \gamma \to A + \overline{A}.$$

As long as the energy of each photon is greater than the required mass energy of the created particles, you're good to go. At a few seconds after the big bang, for instance, the universe was billions of degrees and just energetic enough to create electrons and positrons. Earlier—about a millionth of a second after the big bang—temperatures were in the 10s of trillions Kelvin, and the universe was energetic enough to create (in principle) protons and antiprotons. And so on. Earlier and earlier times were hotter and hotter, and more and more massive particles (and antiparticles) were able to be created "out of thin air" as it were.

Mass and energy can be exchanged in low-energy interactions as well. For instance, the Lyman-alpha transition of hydrogen emits a 10.2 eV photon. The implication is that a ground state hydrogen atom has 10.2 eV less energy than hydrogen in the first excited state.

Here's a more extreme case. Protons and neutrons, as you may know, are made up of three **quarks**—fundamental particles with a charge of +2/3 (up quarks) or −1/3 (down quarks). Protons, for instance, are made of two up quarks and one down quark. The mass energy of a proton is 938 MeV, while the up quarks have masses of 2.2 MeV (with large error bars), and the down quark has a mass of 4.7 MeV* [2]. Add together the mass of a proton, and we get

$$\sum_i m_Q = 2.2MeV + 2.2MeV + 4.7MeV = 9.1MeV \ll M_p.$$

Even more mysteriously, if you swap an up quark for a down, you'd expect the neutron to be 2.5 MeV heavier than the proton, but the difference is only a scant 1.3 MeV.

This difference in mass comes from energy. Just as massive particles can be annihilated to produce energy, the energy of interactions (the strong force) produces a constant sea of energy that is seen as mass via the inverse of Einstein's relation:

$$m = \frac{E}{c^2}.$$

Example: The Lifetime of the Sun

I wanted to do one more detailed numerical example: nuclear fusion in the Sun. While the detailed chemistry in the Sun is a little complicated, the basic reaction is

$$4p \rightarrow 2e^+ + {}^4He + 2\nu_e + 2\gamma, \qquad (5.8)$$

where ν_e are (nearly massless) electron neutrinos and p and He are hydrogen and helium nuclei. In this case, we find that the total masses on the left-hand side of the equation are considerably less than the masses on the right:

- $m_p = 938.3$ MeV
- $m_{He} = 3728.1$ MeV
- $m_e = 0.511$ MeV

$M_i = 3754.2$ MeV and $E_f = 3728.1$ MeV, so

$$\frac{\Delta M}{M_i} = \frac{26.1MeV}{3754.2 \text{ MeV}} = 0.00695.$$

---

*These masses are small enough, relatively, and difficult enough to measure that the uncertainty in both is currently about 10–20 percent.

For every kg of hydrogen fused into helium, about 0.7% of the mass is converted into energy in the form of photons, kinetic energy (heat), and neutrinos.

We can even roughly estimate how long this store of energy will last, but we need to make some extreme simplifying assumptions to do so. The Sun has a total mass of about 2×10^{30} kg, but only 70% of that is hydrogen. Further, only 10% of the gas is hot and dense enough to undergo fusion, and finally, as we've seen, 0.7% of *that* is converted to energy. Thus

$$\Delta E = \left(2 \times 10^{30} \, \text{kg}\right)\left(0.7\right)\left(0.1\right)\left(0.007\right)\left(3 \times 10^{8} \, \text{m/s}\right)^{2} = 8.8 \times 10^{43} \, \text{J}.$$

That certainly seems like a lot, but one way to figure out how much is by looking at how quickly the Sun is going through that energy. The Sun radiates with a luminosity of about

$$L_{\odot} \approx 3.9 \times 10^{26} \, \text{J/s}.$$

Assuming it does so at a constant rate,

$$L = \frac{\Delta E}{\Delta t} \rightarrow \Delta t = \frac{\Delta E}{L},$$

so

$$\Delta t \approx 2.3 \times 10^{17} \, \text{s}$$

$$\approx 7 \, \text{Gy}.$$

There are lots of confounding details, including the fact that the Sun varies a bit in brightness over its lifetime, some of the energy is in the form of neutrinos, new fuel gets convected into the core, and so on. Nevertheless, this result is within 30% of the canonical 10-billion-year main-sequence lifetime of the Sun.

5.3.3 RELATIVISTIC MOMENTUM

We're still missing one crucial component of dynamics: the momentum of relativistic particles. Fortunately, we're able to bootstrap from our derivation of relativistic energy—along with the simple, nonrelativistic form of momentum, $p_{NR} = mv$.

Consider a decaying particle of mass m moving at a nonrelativistic speed v (figure 5.6) into two photons. We've already established that, in the boosted frame, the outgoing photons will have energies

$$E_R \approx (1+v)E_0; \ \ E_L \approx (1-v)E_0.$$

Conservation of momentum requires that

$$p_R + p_L = mv,$$

which is satisfied if

$$p_R = (1+v)\frac{E_0}{c}$$

and similarly for the left-going photon (which includes a minus sign).

More generally,

$$E_\gamma = p_\gamma c, \tag{5.9}$$

which takes on an even simpler form in natural units. In fact, the general relationship between the momentum and energy of light predates the discovery of the photon, going back at least another 40 years to Maxwell himself, [27] who posited that because radiation exerts pressure, it also necessarily carries momentum.

Having drawn the connection to nonrelativistic momentum, going back and computing relativistic momentum is straightforward, again using the configuration in figure 5.6. In natural units (where $|\vec{p}| = E$ for a photon):

$$p = p_R + p_L$$

$$= E_0 \left(\sqrt{\frac{1+v}{1-v}} - \sqrt{\frac{1-v}{1+v}} \right)$$

$$= E_0 \left(2v\gamma \right),$$

or

$$p = mv\gamma, \tag{5.10}$$

which naturally reduces to the familiar nonrelativistic limit in the case of $v \ll 1$.

5.3.4 4-MOMENTUM

We have two wonderful results from our work above:

$$E = m\gamma$$

and

$$p^i = mv^i\gamma.$$

We have four numbers, so we might even combine them into a single 4-vector. We'll call it the **4-momentum**!

$$p^\mu = \begin{pmatrix} E \\ p^1 \\ p^2 \\ p^3 \end{pmatrix} \equiv mU^\mu, \tag{5.11}$$

which has the nice property, as I've included, of being simply m times the 4-velocity. See? There was a good reason for introducing it earlier.

The 4-momentum shows something very important. Energy and momentum are intimately related—related enough that some places insist on calling it "momenergy." (Do so if you must, but I can't condone it.)

This 4-momentum has a lot of nice properties. For instance, because $\mathbf{U} \cdot \mathbf{U} = 1$, we get

$$\mathbf{p} \cdot \mathbf{p} = (m\mathbf{U}) \cdot (m\mathbf{U}) = m^2, \tag{5.12}$$

which can be expanded as

$$E^2 - |\vec{p}|^2 = m^2.$$

It's right triangles all the way down (figure 5.7) As you will occasionally need to expand this in MKS units, it's worth writing it out explicitly:

$$E^2 = (pc)^2 + (mc^2)^2.$$

There's one more math trick we want to do with our 4-momentum. We'll start with a particle at rest. In that case, the 4-momentum is simply

$$p^{\mu(0)} = \begin{pmatrix} m \\ 0 \\ 0 \\ 0 \end{pmatrix}.$$

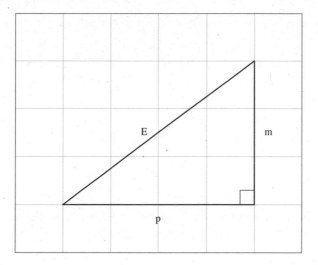

FIGURE 5.7. Energy, momentum, and mass form a right triangle.

We can then apply a boost in the $+x$-direction. As a reminder, this means that to a frame moving at speed v to the left, we can boost all vectors via the transformation matrix

$$\Lambda^{\bar{\mu}}{}_{\mu} = \begin{pmatrix} \gamma & v\gamma & 0 & 0 \\ v\gamma & \gamma & 0 & 0 \\ 0 & 0 & 1 & 0 \\ 0 & 0 & 0 & 1 \end{pmatrix}.$$

Thus,

$$p^{\bar{\mu}} = \begin{pmatrix} \gamma & v\gamma & 0 & 0 \\ v\gamma & \gamma & 0 & 0 \\ 0 & 0 & 1 & 0 \\ 0 & 0 & 0 & 1 \end{pmatrix} \begin{pmatrix} m \\ 0 \\ 0 \\ 0 \end{pmatrix} = \begin{pmatrix} m\gamma \\ mv\gamma \\ 0 \\ 0 \end{pmatrix},$$

exactly as predicted.

The relativistic energy-momentum relationship has some very nice properties. For instance, if we look at a massless particle, we immediately get

$$E = p,$$

as we knew.

But we also can look at the limit of $p \ll E$. In that case,

$$E = \sqrt{m^2 + p^2}$$

$$= m\sqrt{1 + \frac{p^2}{m^2}}$$

$$\approx m\left(1 + \frac{p^2}{2m^2}\right)$$

$$= m + \frac{p^2}{2m},$$

where we used the Taylor series in equation 3.3, the first term is just the mass energy, and the latter is nonrelativistic kinetic energy.

5.3.5 FRAMES OF REFERENCE

One of the recurring themes in relativistic physics is that most observables are in the eye on the beholder—or at the very least, specific to one particular reference frame. While invariant quantities like $\mathbf{p} \cdot \mathbf{p} = m^2$ (eq. 5.12) remain the same regardless of frames, from an observational or experimental perspective, we might be more interested in how a particle's energy—one *particular* component of the 4-momentum—might look in one frame of reference or another.

Consider what happens if we have two observers (or particles, for that matter) moving with respect to some arbitrary third frame. For the moment, we'll just consider their 4-velocities:

$$U_A^\mu = \begin{pmatrix} \gamma_A \\ v_A \gamma_A \end{pmatrix}; \quad U_B^\mu = \begin{pmatrix} \gamma_B \\ v_A \gamma_B \end{pmatrix},$$

where we've written their 4-velocities in $1 + 1$ dimensions only, both for brevity and because we can always rotate our coordinates to align with the axis of their relative motion. What happens when we boost into one particular frame of reference, say, observer B? In that case, we use the coordinate transformation

$$\Lambda^{\bar{\mu}}{}_\mu = \begin{pmatrix} \gamma_B & -v_B \gamma_B \\ -v_B \gamma_B & \gamma_B \end{pmatrix}.$$

In B's frame of reference, observer B is naturally standing still, while A's 4-velocity is

$$U_A^{\bar{\mu}} = \begin{pmatrix} \gamma_B & -v_B\gamma_B \\ -v_B\gamma_B & \gamma_B \end{pmatrix}\begin{pmatrix} \gamma_A \\ v_A\gamma_A \end{pmatrix}$$

$$= \begin{pmatrix} \gamma_A\gamma_B(1-v_Av_B) \\ \gamma_A\gamma_B(v_A-v_B) \end{pmatrix}.$$

This expression is interesting, if a little busy, but we're particularly concerned with the time component, because if we multiply by m_A, we get the 4-momentum, the zeroth component of which is the energy as seen by B:

$$E_{obs} = m_A\gamma_A\gamma_B(1-v_Av_B).$$

Or, more generally,

$$E_{obs} = \mathbf{p}\cdot\mathbf{U}_{observer}, \tag{5.13}$$

which you can quickly show is identical.

While we performed this derivation for massive particles, it works just as well for massless particles. For a photon moving in the $+x$-direction,

$$p_\gamma^\mu = \begin{pmatrix} E \\ E \end{pmatrix},$$

and a receding observer has a 4-velocity

$$U_{obs}^\mu = \begin{pmatrix} \gamma \\ v\gamma \end{pmatrix}.$$

Thus, the apparent photon energy is

$$E_{obs} = \mathbf{p}_\gamma\cdot\mathbf{U}_{obs}$$

$$= E\gamma(1-v)$$

$$= E\frac{1-v}{\sqrt{(1-v)(1+v)}}$$

$$= E\sqrt{\frac{1-v}{1+v}},$$

precisely as we derived in equation 5.1.

5.4 Force*

5.4.1 ACCELERATION

Now that we've established the energy and momentum relationships for relativity, we can start to think a little bit about dynamics. As we saw in Chapter 4, we can get around the whole problem of "relativity requires an inertial reference frame" by looking at the reference frame of an observer moment by moment. We called this the momentarily comoving reference frame (MCRF), and it is astoundingly useful because it allows us to relate external forces to internal sensations.

Ordinary force, such as we might apply externally, is not frame invariant, and, as a result, it's a little more complicated to figure out what happens when we apply an external force to a spaceship. Consider the situation outside the spaceship. A handy relation can be computed via

$$\frac{d\gamma}{dv} = \frac{v}{(1-v^2)^{3/2}} \tag{5.14}$$
$$= v\gamma^3.$$

We can use the calculation to figure out what Newton's second law really tells us in special relativity:

$$\frac{dp}{dt} = \frac{d(mv\gamma)}{dv}\frac{dv}{dt}$$
$$= ma\left[\gamma + v\frac{d\gamma}{dv}\right]$$
$$= ma\left[\gamma + v^2\gamma^3\right]$$
$$F = ma\gamma^3,$$

which is the relativistic version of Newton's second law (albeit still written in terms of 3-vectors). Because $\lim_{v\to 0}\gamma = 1$, this calculation yields the classic relation taught in high schools.

We can also ask how this force feels to someone inside the spaceship. Consider a short time step during which the spaceship speeds up from v to $v + a\Delta t$, where

$$a = \frac{F}{m\gamma^3}.$$

In a frame moving at speed v, then, at first $\bar{u} = v$ corresponds to $u_1 = 0$. That is, at some instant, a person is at rest with respect to the spaceship. A short time later,

$$u_2 = \frac{(v + a\Delta t) - v}{1 - (v + a\Delta t)v},$$

which, to first order in Δt is

$$u_2 \approx a\gamma^2 \Delta t,$$

so

$$g = \frac{\Delta u}{\Delta \tau} = \frac{u_2 - u_1}{\Delta \tau}$$
$$= \frac{a\Delta t\gamma^2}{\Delta t/\gamma}$$
$$= \frac{F}{m}.$$

Holy moly!

So the artificial gravity is $g = F/m$, regardless of how fast the spaceship is already going.

5.4.2 RELATIVISTIC ROCKET SCIENCE

Having developed the equations for an accelerating, relativistic particle, it's time to do some rocket science. As we've seen, if we want the inhabitant of the spaceship to experience a constant acceleration (thus, artificial gravity), we simply apply a constant force $F = mg$.

We can solve the parameters of the trip (how long it takes according to the inhabitant of the spaceship, according to Earth, according to the speed of the spaceship) parametrically in terms of the distance traversed. We begin by using the work-energy relation

$$F \, dx = dE$$
$$mg \, dx = md\gamma,$$

or

$$\gamma = 1 + gx,$$

where the boundary condition of $\gamma = 1$ at $x = 0$ comes from the assumption of starting at rest.

We may then invert the γ relation to solve for the speed of the spaceship as a function of distance:

$$v = \sqrt{1 - \frac{1}{\gamma^2}}$$
$$= \sqrt{1 - \frac{1}{(1 + gx)^2}}.$$

Finally, we can relate the coordinate time to distance via the speed of the spaceship:

$$dt = \frac{dx}{v},$$

so

$$t = \int \frac{dx}{\sqrt{1 - \frac{1}{(1+gx)^2}}}$$

$$= \sqrt{\frac{x(2+gx)}{g}}. \tag{5.15}$$

We can also figure out how much proper time passes for the inhabitant of the spaceship by calculating the proper time in each MCRF and adding them up. That is,

$$\tau = \int \frac{dt}{\gamma}$$

from the beginning of the trip to the end of the trip. Using the relations from above:

$$\tau = \int \frac{dt}{\gamma}$$

$$= \int \frac{dx}{\sqrt{1 - \frac{1}{(1+gx)^2}}(1+gx)}$$

$$= \frac{\ln\left(1 + gx + \sqrt{gx(gx+2)}\right)}{g}. \tag{5.16}$$

The divergence of the different measures of time can be seen in figure 5.8. After approximately a year, the ship starts moving at relativistic speeds, and thus makes a nearly 45-degree angle in the spacetime diagram. As that happens, the increase in proper time within the spaceship grinds nearly to a halt.

As a practical matter, relativistic acceleration could be used to make an arbitrarily distant trip in a very short amount of time from the passenger perspective. For instance, it's approximately 25,000 light-years [20] to the center of the galaxy. A human colony would take about 12,500 years—as seen on Earth—to accelerate halfway there, and the same to decelerate. On the ship, however, a little under 20 years would pass. You could do it in a human lifetime!

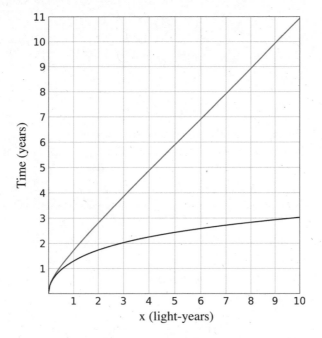

FIGURE 5.8. The spaceship's clock (black, lower curve) and the Earth clock (gray, upper curve) for a spaceship accelerating at a constant internal acceleration g.

While travel times to other stars would be very short for the crew, the amount of fuel used would be enormous. Let's suppose that you have a matter-antimatter engine (half of each). Starting from rest, if an amount of fuel dM is converted to energy, $dE = dM$ energy is created. If it's sent out the back of the ship, it will carry a momentum of

$$p_\gamma = -dE = -dM.$$

By Newton's third law (in the nonrelativistic frame of the spaceship itself),

$$Mdu = -dM,$$

or

$$M\frac{du}{d\tau} = Mg = -\frac{dM}{d\tau}.$$

Thus,

$$gd\tau = -\frac{dM}{M}$$

or (here's your rocket science!)

$$g\tau = \ln\left(\frac{M_f}{M}\right),\tag{5.17}$$

which yields

$$M = M_f \exp(g\tau).\tag{5.18}$$

Here, M_f is the total mass of cargo (crew, air, the spaceship itself), while M is the required total mass the ship needs to start with (everything in M_f plus the fuel). So, for instance, for a four-year trip (as experienced by the astronauts), you need to start with 61.6 times as much mass as you will have at the end of the trip. Roughly 98.3% of the mass of the spaceship will be matter and antimatter.

Given that even the relatively modest Apollo 11 lander had a mass of about 50 tons, that means we'd need to create and safely store about 1,500 tons of antimatter to make the trips we're talking about. Suffice it to say, that's well beyond our technology at present.

Looking Forward

In this chapter, we took a good hard look at relativistic dynamics of individual particles and finally derived $E = mc^2$. This is not all there is to relativistic physics, of course, and a big missing piece is what happens when we have many particles—enough to form a continuous fluid. We will address that central question in Chapter 6, and our discussion will provide a foundation for understanding and anticipating how the sources of electricity and gravity will ultimately show up in those theories.

5.5   Problems

1. A helium laser emits light at 632.8 nm in the rest frame of the laser.
   (a) Suppose the laser source is pushed toward you at a speed of 30 m/s. By what Doppler shift, z, would the light be shifted?
   (b) How would your answer change if you moved toward the source instead of the other way around?
   (c) Suppose the source moves away from you at a speed of $v = 0.5$. At what wavelength would you observe the light? What is the corresponding Doppler shift?
   (d) Suppose it moves toward you at $v = 0.5$. What is the observed wavelength and corresponding Doppler shift?
2. A ship is on the way to alpha Centauri (distance = 4 ly according to Earth) at a speed of $v = 0.6$. In the spaceship frame, what is the distance to the star?

3. As a way of easing into mass, momentum, and energy, I'd like us to return to natural units. All three (mass, momentum, and energy) can be expressed in energy units. Compute the following in eV:

   (a) The rest mass of an electron

   (b) The kinetic energy of an electron at a speed of 2.2×10^6 m/s (roughly the orbital "speed" of an electron in a hydrogen atom; you can use the nonrelativistic energy)

   (c) The momentum of an electron at a speed of 2.2×10^6 m/s

4. There are many ways of deriving the relativistic momentum relation. One particularly clean one involves starting with the fundamental definition of impulse, $\Delta p = F \Delta t$, and relating it to work, $\Delta E = F \Delta x$. The two equations can be combined to yield a general relation

$$\frac{dE}{dv} = v \frac{dp}{dv}.$$

Although "show that" problems generally result in endless algebra, we'll make an exception in this case. Show that the equation for relativistic energy (eq. 5.6) and the relation above lead inexorably to the equation for relativistic momentum (eq. 5.10).

5. Consider a 10,000 kg spaceship moving at $v = 0.8$.

   (a) What is the *mass* energy of the ship?

   (b) What is the *total* energy of the ship?

   (c) What is the *kinetic* energy of the ship?

   (d) Suppose you did the calculation by (wrongly) using the nonrelativistic kinetic energy formula. What do you get then?

   (e) What is the momentum of the spaceship? (Use natural units.)

   (f) What is the same momentum in MKS units?

6. Radium-226 has a half life about 1600 years, after which it decays into radon 222 and a so-called alpha particle (a helium nucleus):

$$^{226}Ra \rightarrow {}^{222}Rn + {}^{4}He.$$

The three components have the following masses:

   - $m_{Ra\text{-}226} = 210.5427$ GeV (226.0254 amu)
   - $m_{Rn\text{-}222} = 206.8094$ GeV (222.0176 amu)
   - $m_{\alpha} = 3.7284$ GeV (4.0026 amu)

   (a) How much energy is released in the decay?

   (b) Assume that all of the energy goes into the kinetic energy of the alpha particle. What is the corresponding momentum?

   (c) How fast is the alpha particle emitted?

7. Let's do a similar decay:

$$A \rightarrow B + C$$

with

- $m_A = 10\,\text{MeV}$
- $m_B = 4.9\,\text{MeV}$
- $m_C = 0.1\,\text{MeV}$

   (a) How much energy is released in the decay?

   (b) Assume that all of the energy goes into the kinetic energy of particle C. What is the corresponding momentum?

   (c) How fast is particle C emitted?

8. A proton has a mass of 938 MeV in the Large Hadron Collider and a speed of $v = 0.9999994$.

   (a) What is the 4-momentum of an individual proton?

   (b) As a practical matter, individual protons slam into one another from opposite directions (that's why it's called a collider). What is the energy of a proton as measured in the frame of reference of another?

   (c) Suppose you wanted to fly comfortably to the other side of the galaxy, about 50,000 light-years away. If you accelerated at g for half the trip and then decelerated at g for the second half, how long would it take you, according to your own clocks? How much time would have passed in the rest of the galaxy?

FURTHER READINGS

- David Bodanis, $E = mc^2$, *A Biography of the World's Most Famous Equation.* (Walker, 2000). A very readable work that serves a good companion, with excellent historical context, to the numerical work that we're doing in this chapter (and throughout the book).

- Stephen Boughn, "Fritz Hasenöhrl and $E = mc^2$," *The European Physical Journal H* 38, no. 2 (2013): 261–78. I've noted earlier that this is not a history text, but I've clearly made a choice in ascribing unambiguous precedence to Einstein in terms of the discovery of $E = mc^2$. Others, including Hasenöhrl, studied black-body radiation and found that radiation must carry away mass in roughly (up to a pre-factor) E/c^2.

- Sidney Coleman, *Sidney Coleman's Lectures on Relativity*, ed. David J. Griffiths, David Derbes, and Richard B. Sohn (Cambridge University Press, 2022). Coleman's perspective provides an important insight into the workings of relativity. The first half of the book focuses on special relativity, including a strong focus on the mass-energy relation addressed in this chapter.

- Richard Feynman, Matthew Sands, and Robert B. Leighton, *The Feynman Lectures on Physics* (Addison-Wesley, 1963). The entire set of Feynman lectures are classics, but Lecture 15 is especially relevant to the current discussion. Readers should be aware that Feynman treats mass itself as relativistic, as opposed to the relativistic energy.

6

Fluids

Leonhard Euler (1707–1783). Euler developed much of the foundational research in fluid dynamics (as well as classical mechanics and pure mathematics), and Euler's equations are named in his honor. As we shall see, our tensor notation allows us to arrive at some of the same results based on Lorentz invariance.

6.1 Dust

The bulk of our work so far has focused on the dynamics of individual particles: how energy, momentum, mass, and velocity all relate to one another and how discrete objects move through spacetime. But for many systems of interest, there are lots of particles—so many particles, in fact, that it stops being worthwhile to talk about what this or that particle is doing, but rather, we need to think instead about their ensemble behavior. This is true, for instance, for the air in the atmosphere—you could, of course, try to model every molecule individually, but I wouldn't recommend it.

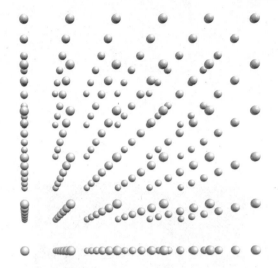

FIGURE 6.1. A simple, stationary dust. In this case, the density is presumed to be uniform throughout all of space and time. Kind of boring physics, if you ask me.

Instead, we're going to take a lesson from calculus and deal with the limit of continuous fields. Fields can take on many forms. They can refer to the density field of a fluid or the electromagnetic or gravitational fields (the topics of the next two chapters, respectively). They can refer to quantum mechanical waves. They can refer to anything that varies continuously in space and time and on which you can take derivatives. You can't take a derivative of an electron, but you can take derivatives of a potential energy field.

Our focus in this chapter is a bit more specific. We're looking on collections of particles known as **fluids**. Fluids, for our purposes, are substances that can be distorted in shape but have a continuous density of … something. That something could be electrical charge, energy, spin, linear or angular momentum, or all of the above. Fluids also serve as the sources for electromagnetism and gravity. They also, incidentally, play very nicely within the framework of special relativity.

We'll build up our picture of fluids by way of example. The simplest possible case is what is commonly called a **dust**. Dusts are more or less what they sound like: many, many particles suspended or moving through space but otherwise not interacting with each other at all.

Imagine a lattice of particles of mass m, charge q, and otherwise (for our first foray) at rest (figure 6.1). If each particle is separated from its neighbor by Δx in each direction, then the number density of the region is

$$n_0 = \frac{1}{\Delta x^3},$$

where we use the subscript "0" to indicate that we're specifically looking at the rest frame of the fluid. This lattice starts to look *continuous* once the separations become very small. In air, for instance, typical separations between molecules are only a few nanometers, which is close enough to treat air as continuous, even though, strictly speaking, it isn't.

Once we know the number density, computing quantities like mass density or electrical charge density is a breeze:

$$\rho_{M,0} = mn_0; \quad \rho_{E,0} = qn_0. \tag{6.1}$$

Normally, we wouldn't bother with all of the subscripts because the context will likely make the type of density clear. In this case, though, the pedantry will help to defuse any ambiguity.

Our starting point has us making all regions of our dust equally spaced, but that need not be so. You can certainly have regions of high density, lower density, and gradients between the two. Likewise, because all of these individual charged particles are stationary, there is no electrical current, but as we'll see, we can easily generate one.

6.2   4-Current

Consider what happens when we look at the particles in a boosted frame. For concreteness, suppose we boost everything by a large velocity, v, in the x-direction. The density of the particles is still uniform, but in the boosted frame, they are closer together (figure 6.2).

In this case, the charge density is increased to $\rho_E = \gamma \rho_{E,0}$, and likewise, since the current of each particle is qv, the current density in the x-direction is $J^x = qv\gamma n_0$. Putting this all together, it seems as though we can describe the charge and current density as a single **4-current**:

$$J^\mu = \begin{pmatrix} \rho \\ J^x \\ J^y \\ J^z \end{pmatrix}. \tag{6.2}$$

The density, ρ, is how much charge you have in one place per unit volume, while J^x (for instance) tells you the density of current flow in the x-direction.

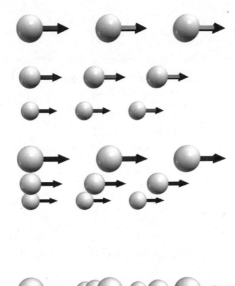

FIGURE 6.2. A dust in a frame boosted in the +x-direction.

In the rest frame of the fluid, this is simply $\left(\rho_0\ 0\ 0\ 0\right)$, which can then be boosted (for instance, in the x-direction) to

$$J^{\bar{\mu}} = \Lambda^{\bar{\mu}}{}_{\mu} J^{\mu}$$

$$= \begin{pmatrix} \gamma\rho_0 \\ v\gamma\rho_0 \\ 0 \\ 0 \end{pmatrix}. \tag{6.3}$$

Equations 6.2 and 6.3 can then be directly compared to interpret how the density and current relate to the rest-frame density.

Looking at the 4-current, you might be tempted to think that it's identical to the 4-momentum or any other 4-vector we've seen so far. Well, it is and it isn't. You can still apply all of the tensor rules we've seen so far. You can use the metric to lower an index or contract the 4-current with a 1-form or whatever you like. But what makes it different is that, while the 4-momentum is defined for a particular particle, when we're talking about fields (the 4-current), we mean that it has a quantity that is defined, and may differ, at all points in space and time. For a particle property, we generally only think of 4-momenta as

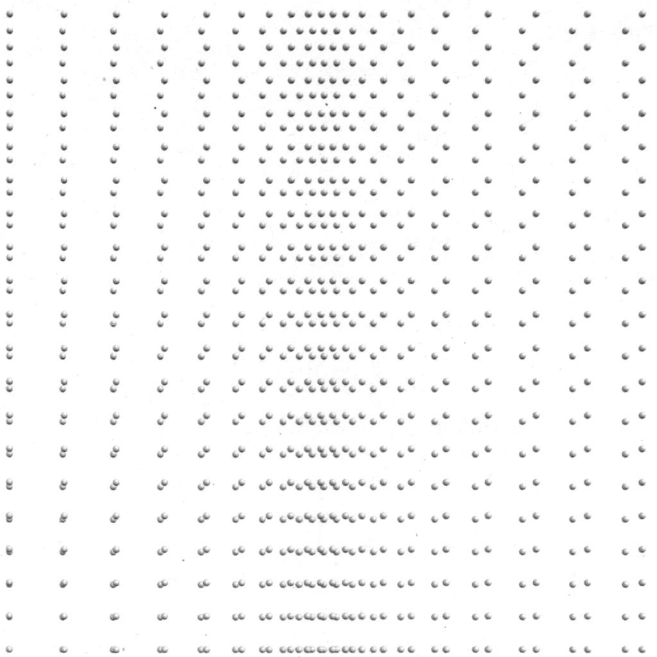

FIGURE 6.3. A dust of nonuniform density. To turn this into a continuous density field, we'd have to smooth over some scale, but no matter the scale, it is clear by visual inspection that the density is higher toward the center (in the x-direction) than near the edges.

changing in time. Changes in space and time are going to be very important when we think about currents.

6.2.1 DERIVATIVES

To envision a concrete example of change, consider the stationary dust shown in figure 6.3. We determined earlier that space and time are similar units, but we're now going to see just how similar they are.

As with all physical calculations of change, we're going to take derivatives – derivatives in time and derivatives in space. To make things look cleaner, we're going to introduce one more tensor rule to the previous five.

6. Derivatives introduce another downstairs index.

In tensor notation, derivatives have an especially nice form:

$$\partial_\mu \phi = \frac{\partial \phi}{\partial x^\mu}. \tag{6.4}$$

In other words, we write the derivative as a "downstairs" index, and thus (using only 3-vector notation), we can take the **divergence** of a vector in a very simple way:

$$\partial_i A^i = \partial_1 A^1 + \partial_2 A^2 + \partial_3 A^3$$

$$= \frac{\partial A^1}{\partial x} + \frac{\partial A^2}{\partial y} + \frac{\partial A^3}{\partial z}$$

$$= \nabla \cdot \vec{A}.$$

The downstairs and upstairs indices "cancel" by the Einstein summation convention, and we're left with a scalar field.

Example: The 4-Divergence of Position

Consider the spacetime position vector

$$x^\mu = \begin{pmatrix} t \\ x \\ y \\ z \end{pmatrix}.$$

What is the 4-divergence of the position?

Solution
The divergence may be written

$$\partial_\mu x^\mu = \frac{\partial t}{\partial t} + \frac{\partial x}{\partial x} + \frac{\partial y}{\partial y} + \frac{\partial z}{\partial z} = 4.$$

6.2.2 WHY "UPSTAIRS" AND "DOWNSTAIRS" INDICES
        ARE DIFFERENT

Early on, I made a big deal about the fact that you can't simply mix and match upstairs and downstairs indices—that an object with an upstairs index, A^i is a list of the components of a vector, while one with a downstairs index is a 1-form.

To make this clear, consider a simple, one-dimensional universe in which we're measuring the temperature as a function of position:

$$T = T_0 + \alpha x.$$

In this particular case, over a displacement of $\Delta x = 1$, there is an increase in temperature of α. Or, to put it another way,

$$\partial_1 T = \alpha.$$

Now, let's suppose that we converted from meters to centimeters. In that case, our transformation "matrix" (it's only got one element) is

$$\Lambda^{\bar{1}}_{\ 1} = 100,$$

and so the displacement in the new frame is

$$\Delta x^{\bar{1}} = \Lambda^{\bar{1}}_{\ 1} \Delta x^1 = 100.$$

The number (obviously) increased by a factor of 100!

But what of the derivative, the 1-form?

$$\partial_{\bar{1}} T = \frac{\Delta T}{\Delta x^{\bar{1}}} = \frac{\alpha}{100}$$

The vector increases by 100, but the 1-form decreases by 100. But—and here's the important point—

$$\partial_i T \Delta x^i$$

is a constant regardless of frame.

6.2.3 CONSERVATION LAWS

In a letter to Leonhard Euler, eighteenth-century Russian physicist Mikhail Lomonosov proposed conservation of mass as a scientific theory. We now know that conservation of mass is, strictly speaking, wrong—that is one of the central results of special relativity—but for most purposes, it works well enough. Euler is generally credited with writing down the first set of equations describing the conservation of mass. In modern notation, we'd write the first of Euler's equations, the **continuity equation**, as

$$\dot{\rho} + \nabla \cdot \vec{J} = 0. \tag{6.5}$$

While Euler originally developed this formula for mass conservation, we'll find a nearly identical relation for conservation of charge and energy.

It's worth spending a moment to think about what equation 6.5 *means*, even in nonrelativistic physics. The mass density, ρ, in a particular region of spacetime can change, but for it to increase (for instance), more particles need to flow in than flow out. For nonrelativistic mass conservation, the flow is given by

$$\vec{J}_M = \frac{d(\rho_M \vec{v})}{dt},$$

and we can see a simple one-dimensional example of this effect in figure 6.4.

$$\vec{J}$$

FIGURE 6.4. The flow of some quantity (mass, in Euler's thinking) through a region of space. Darker areas indicate higher density. The gradient is negative (decreasing current), so $\frac{dJ^x}{dx} < 0$, which means that the mass in the "clump" is increasing.

The continuity equation can be made clearer by integrating the density over some (any) volume,

$$\int \dot{\rho} d^3 x = \dot{M},$$

to yield the total mass within that volume. Likewise, the divergence theorem in calculus gives the handy relation

$$\int d^3 x \nabla \cdot \vec{J} = \oiint dS \vec{J} \cdot \hat{n}, \tag{6.6}$$

where $\hat{n}$ is the unit vector pointing out of the surface. Applying the divergence theorem to the continuity equation, we get

$$\dot{M} = - \oiint dS \vec{J} \cdot \hat{n}. \tag{6.7}$$

For any arbitrary volume, the change in mass is given by the net flow in or out of the volume.

While mass isn't conserved, the form of Euler's continuity equation is useful for all sorts of things, including our electrical 4-current. But what's more, it comes out naturally. By "naturally," I mean that the benefit to all of our hard work on tensor rules is that when we can write an equation where all of the indices cancel, if it's true in one Lorentz frame, then it's true in all. For example, consider the equation

$$\partial_\mu J^\mu = 0. \tag{6.8}$$

Since all indices cancel, the term is a scalar. What's more, not only is the expression Lorentz invariant—and thus a possibility for a physical law – it just so happens to be identical to equation 6.5.

This is a bigger deal than you might realize at first. One of the central ideas of relativity is that it represents the rules of the road for how our universe functions. But there's another way of saying that. Because we have been incredibly pedantic regarding the rules for constructing tensor equations (and how coordinates work), any equation that satisfies the tensor rules (summarized in Section A.2 in the Appendix) automatically satisfies Einstein's postulates— or, to put it another way, automatically makes them Lorentz invariant. So if you're trying to construct an equation that involves taking a derivative of a current, equation 6.8 is pretty much the only one you can write down that fits the bill. In other words, once we'd generated a "good" electromagnetic 4-current, conservation of electric charge immediately gives rise to an equation like equation 6.8.

Example: Current Contour

Consider a 1+1 dimensional current

$$j^\mu = \begin{pmatrix} \rho \\ \alpha x \end{pmatrix},$$

where α is some positive constant. This relation is only valid over some range. What is the instantaneous change in density at the origin?

Solution
Applying the continuity equation,

$$\partial_\mu j^\mu = \dot\rho + \alpha = 0,$$

and thus

$$\dot\rho = -\alpha.$$

Less material flows in than out; thus the density is decreasing.

6.2.4 WHAT HAPPENS IF THE METRIC ISN'T FLAT?*

Because the physics of fluids relies heavily on the divergence $\partial_\mu V^\mu$, we've swept a not-so-minor detail under the rug: this simplistic form depends on using either the Cartesian or Minkowski metrics (in space and spacetime, respectively). But things become markedly more difficult in different coordinate systems. One of the mathematical challenges of working in general relativity is that we can't just apply a simple, fixed metric to the universe.

To get at least a sense of the challenge, consider generalizing the divergence of a vector in ordinary space:

$$\nabla \cdot \vec{A} = \text{scalar}.$$

The divergence of a vector (whether in space or spacetime) is simply a scalar. Scalars are incredibly useful. They are, indeed, the reason that we've been so keen on contracting indices to produce them. For instance, a particle's 4-momentum has the relation $p \cdot p = m^2$. This is true independent of frame, and just as true independent of coordinate system. It's just a number, in other words.

But now, consider what would happen if we try to evaluate the divergence in curvilinear coordinates, including cylindrical or spherical or one of their more complicated cousins that arise from general relativity. The curved coordinate system is denoted with a bar. Thus,

$$\nabla \cdot \vec{A} = \partial_i A^i$$

$$= \left(\Lambda^{\bar{i}}_{\ i} \partial_{\bar{i}} \right) \left(\Lambda^i_{\ \bar{j}} A^{\bar{j}} \right),$$

where, as a reminder, the Λ matrices are transformations. It's been a while, but look back at, for instance, equation 1.11, in which we computed the transformation between Cartesian and polar coordinates and found that Λ was an explicit function of r and θ. In that case, there are nonzero derivatives of both $A^{\bar{j}}$ and the transformation matrix. Continuing the expansion from above,

$$\nabla \cdot \vec{A} = \Lambda^{\bar{i}}_{\ i} \Lambda^i_{\ \bar{j}} \partial_{\bar{i}} A^{\bar{j}} + A^{\bar{j}} \left(\Lambda^{\bar{i}}_{\ i} \partial_{\bar{i}} \Lambda^i_{\ \bar{j}} \right)$$

$$= \partial_{\bar{i}} A^{\bar{i}} + A^{\bar{j}} \left(\Lambda^{\bar{i}}_{\ i} \partial_{\bar{i}} \Lambda^i_{\ \bar{j}} \right).$$

The first term looks exactly as you'd expect (and identical to what we saw in Cartesian coordinates), but the latter is new. The term in the parentheses is known as the "Christoffel symbol," and it essentially describes the variation in coordinate axes. I introduce it not because I want you to compute it for any particular coordinate system—indeed, it's in a particularly ugly and tedious form to try to do so—but because I wanted to justify why we are focusing on only one coordinate system in this book. It allows us to avoid the complications of Christoffel symbols—a luxury that general relativists don't have.

6.3   The Stress-Energy Tensor

While the original formulation of the continuity equation was based on the (erroneous) belief in the conservation of mass, we now know that energy is the conserved quantity. It's even the first of Rudolf Clausius's laws of thermodynamics. A half-century later, in 1917, Emmy Noether showed that conservation of energy was a natural consequence of the time invariance of physical laws.

If energy is conserved over time, we need a way of representing it in fluid form that is meaningful in the context of Lorentz invariance. But a 4-vector current (eq. 6.3) won't do the trick. Here's why: when we boost a dust, two things happen. First, in exactly the same fashion as with charge density, the separation between particles will contract by a factor of γ. But in addition, the energy of every individual particle will increase by a factor of γ. Thus, the energy density in the boosted frame increases by γ^2.

In order to effectively describe energy density in relativity, we need a two-index (rank-2) tensor, known as the **stress-energy tensor**. As with the current density, where the zero component corresponded to density, we're going to suppose that $T^{00} = \rho$ for the stress-energy tensor. At rest, a dust can be written as

$$T^{\mu\nu(0)} = \begin{pmatrix} \rho_{M,0} & 0 & 0 & 0 \\ 0 & 0 & 0 & 0 \\ 0 & 0 & 0 & 0 \\ 0 & 0 & 0 & 0 \end{pmatrix}.$$

We can then boost this in any arbitrary direction. For instance, boosting into the positive x-direction,

$$T^{\mu\nu} = \Lambda^{\mu}_{\ \overline{\mu}} \Lambda^{\nu}_{\ \overline{\nu}} T^{\overline{\mu\nu}(0)}$$

$$= \rho_{M,0} \begin{pmatrix} \gamma^2 & v\gamma & 0 & 0 \\ v\gamma & v^2\gamma^2 & 0 & 0 \\ 0 & 0 & 0 & 0 \\ 0 & 0 & 0 & 0 \end{pmatrix},$$

where, as expected, the density is boosted by a factor of γ^2. But there's more. Since this particular case represents a fluid flowing in the positive x-direction, T^{10} represents the momentum density in the x-direction while T^{01} (numerically identical) corresponds to the energy flow in the x-direction.

We could imagine constructing a simple, locally stationary fluid by adding equal components of boosted fluids in all directions. Upon doing so, all but the diagonal components vanish, leaving us with

$$T^{\mu\nu} = \begin{pmatrix} \rho & 0 & 0 & 0 \\ 0 & P & 0 & 0 \\ 0 & 0 & P & 0 \\ 0 & 0 & 0 & P \end{pmatrix}. \tag{6.9}$$

Or, allowing for velocity flows,[*]

$$T^{\mu\nu} = \begin{pmatrix} \text{Density} & x\text{-Momentum} & y\text{-Momentum} & z\text{-Momentum} \\ \text{Energy Flow-}x & \text{Pressure} & x\text{-}y\text{ Shear} & x\text{-}z\text{ Shear} \\ \text{Energy Flow-}y & x\text{-}y\text{ Shear} & \text{Pressure} & y\text{-}z\text{ Shear} \\ \text{Energy Flow-}z & x\text{-}z\text{ Shear} & y\text{-}z\text{ Shear} & \text{Pressure} \end{pmatrix}. \tag{6.10}$$

The primary purpose of developing the stress-energy tensor is to describe mathematically what we mean when we talk about conservation of energy, for instance. We would ideally like something similar to equation 6.8, which we can get in the form of

$$\partial_\mu T^{\mu\nu} = S^\nu.$$

We don't have just one equation; we have four! The source term on the right is to be determined, but for an isolated system, we can guess what it is: zero. No change is essentially what we *mean* by a conservation law. Thus, for $\nu = 0$,

$$\partial_\mu T^{\mu 0} = 0 \tag{6.11}$$

is our continuity equation, but for energy.[†] Likewise, for spacelike values of ν,

$$\partial_\mu T^{\mu i} = 0 \tag{6.12}$$

are three equations representing conservation of linear momentum in each of the three spatial dimensions.

We'll spend the last two chapters thinking a little about how these fluids might serve as sources and, knowing that they are conserved, how that can constrain what laws of physics are possible.

----

[*] This description is primarily just for your edification. We won't be doing major manipulations of the stress-energy tensor, but if you want to give a name to individual elements, these are the ones to use.

[†] Expand it explicitly, if you don't believe me.

Looking Forward

In this chapter, we developed a nice, clean way of writing down charge and current densities as a single tensor and did the same thing for energy density, momentum density, pressure, and stress. But then we made an important breakthrough about the rules of physics under the constraints of relativity: if the tensor rules are satisfied, then so is Einstein. Adhering to tensor rules doesn't mean that a proposed equation is correct, but violating them *does* mean that it's incorrect. In the next two chapters, we're going to see how the fluids that we developed in this chapter fit into physical theories, starting with Maxwell's equations of electromagnetism.

6.4 Problems

1. Consider a current 4-vector flowing in the x-direction plane:

$$J^\mu(x) = \begin{pmatrix} \rho(t) \\ A\cos(kx - \omega t) \\ 0 \\ 0 \end{pmatrix},$$

   where the amplitude of the current, A; the wavenumber, k; and the frequency, ω, are all positive constants. Using the continuity equation, solve for $\rho(t)$. As with all integrals, the solution may include an additive constant.

2. Many applications of relativistic waves and fluids use Euler's exponential relation $e^{i\theta} = \cos\theta + i\sin\theta$ as a basis to describe a fluid. The expression might be

$$\psi = \psi_0 e^{i\mathbf{p}\cdot x},$$

   where $\mathbf{p}$ is some unknown 4-momentum.

   (a) Compute $\partial^\mu \psi$.

   (b) Use the continuity equation to determine if your answer from part (a) is a conserved current (ignoring the messiness of the fact that the numbers are complex). If so, what are the constraints on $\mathbf{p}$?

3. Consider a 1+1 dimensional dust moving at nonrelativistic speeds. That is,

$$T^{\mu\nu} = \begin{pmatrix} \rho & v\rho \\ v\rho & v^2\rho \end{pmatrix},$$

   where both ρ and v are allowed to vary in position and time.

   (a) Expand $\partial_\mu T^{\mu 0} = 0$ explicitly in terms of spatial and time derivatives.

   (b) Expand $\partial_\mu T^{\mu 1} = 0$ explicitly in terms of spatial and time derivatives.

   (c) Now, assume that $\rho = \rho_0(1 + \delta)$, where δ is a fractional density contrast. It is assumed to be small, such that $\delta^2 \approx \delta v \approx 0$. Expand the first (continuity) equation in linear order of δ and v.

(d) The pressure of a fluid can be expressed as ρv^2. With that in mind, what is the linearized version of $\partial_\mu T^{\mu 1} = 0$?

4. The stress-energy tensor is key to gravity; it is central in many cosmological predictions. In particular, a useful simplifying assumption in many cosmological applications is that the flow of matter and energy is near zero (i.e., that you can use eq. 6.9 for the stress-energy tensor) and that $w = P/\rho$ is a constant.

(a) Starting with equation 6.9, boost the tensor in the x-direction by an amount v. Your answer should include only factors of ρ, w, v, and γ.

(b) Ordinary matter and dark matter have $w \approx 0$. What are the elements of the boosted stress-energy tensor at speed v?

(c) You may also have heard of a component of the universe called "dark energy," which has a surprising equation of state: $w = -1$. Compute the boosted stress-energy tensor for dark energy, being sure to simplify as much as possible.

FURTHER READINGS

- Rutherford Aris, *Vectors, Tensors, and the Basic Equations of Fluid Mechanics,* (Dover, 1990). Many undergraduate physics sequences teach fluids in fits and starts, with the presumption that students will pick up what they need from relativity, mechanics, cosmology, or other courses as needed. This book can be used to address that deficit. It is very readable, and while its tensor notation differs somewhat from that used here, you should still be able to readily translate between the two.

- John Peacock, *Cosmological Physics.* (Cambridge University Press, 1998). Many students will first encounter the stress-energy tensor and relativistic fluid flows in the context of structure growth in cosmology. Peacock does an excellent job of outlining the basic approach to using these objects in cosmological contexts.

- Bernard Schutz, *A First Course in General Relativity*, 3rd ed. (Cambridge University Press, 2022). This book has an outstanding discussion of currents and the stress-energy tensor in Chapter 4.

Electromagnetism

James Clerk Maxwell (1831–1879). Although Maxwell didn't discover the laws that bear his name, he did collect and unify them. As we will see, Maxwell's laws give hints of the unity not only between electricity and magnetism but also between space and time.

We've been going about the development of relativity in a completely ahistorical way. Einstein's first paper on the subject wasn't about mass-energy equivalence but rather "On the Electrodynamics of Moving Bodies," [9] a topic we've more or less ignored until now. But that is about to change. As we will soon see, not only does electromagnetism allow us to exploit the manipulations we've developed in space and time, it also provides an outstanding example of what a relativistic theory *should* look like.

7.1 Moving Charges

We start with a straightforward riddle, one that's prompted by Einstein's first postulate: an observer shouldn't be able to tell if they are moving or standing

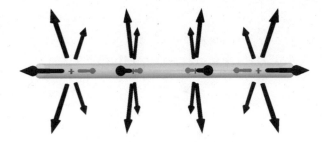

FIGURE 7.1. The electric field generated by an infinite wire with constant electrical charge. The field lines point radially outward from the wire, and, from Gauss's law, we find that, as you double the distance, the field drops off as $1/r$. Just to allay any confusion, note that the electric field from a point charge—as opposed to a wire—will drop off as $1/r^2$, the same as gravity.

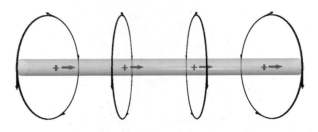

FIGURE 7.2. This wire is similar to figure 7.1 but has a current running through it generating a magnetic field along with the electric field seen in the figure 7.1. If you point your right-hand thumb in the direction of the flow of the current, your remaining fingers will curl around the wire in the direction of the magnetic field. Above the wire, for instance, the field lines come out of the page. As with the electric field, the magnetic field scales as $1/r$ with distance from the wire.

still. Take a wire of infinite length with a fixed charge per unit length λ. The wire creates an electric field, or E-field, emanating outward as shown in (figure 7.1).

The system need not be static. If the individual electrons are moving at a speed of v, there is a current in the wire: $I = \lambda v$. Taking the same wire from figure 7.1 and moving the charges gives us a magnetic field, or B-field (figure 7.2).

Einstein's critical realization was that by looking at figure 7.1 in a moving frame of reference, the charges will appear to move, and thus, if the laws of physics really are invariant, we'd expect to see a magnetic field, as in figure 7.2. That's it! An electric field becomes an electric *and* a magnetic field just by changing frames of reference.

Now consider a second, identical charged wire running parallel to the first. In the rest frame of the two wires, the two (having like charges) will repel each other electrically. However, in the boosted frame, each of the wires carries a current. The current from one will be attracted magnetically by the field from

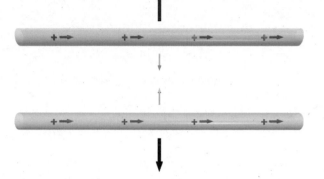

FIGURE 7.3. Two parallel, charged wires with currents running in the same direction will attract each other magnetically while repelling each other (much more strongly) electrically.

the other (figure 7.3). While the magnetic attraction is less than the electrical repulsion, it remains the case that, absent relativity, the force between the two wires seems to differ between frames.

We know (because we worked out the dynamical implications of relativity already) that force is not a Lorentz-invariant quantity. But the rest of this chapter will be devoted to the question of how the electromagnetic fields themselves transform.

7.2 Maxwell's Equations

I don't anticipate that you've had an advanced electromagnetism course, so I've provided a brief, annotated introduction to the topic with the modest goal of gleaning hints as to similarities and differences between the electric and magnetic fields.

Maxwell's equations, the four core equations in electromagnetism, provide rules for how currents and charges give rise to electromagnetic fields. I'm not going to derive them; just state them:

1. **Gauss's Law**

   Gauss's law tells us how charge creates an electric field. It's the equation that shows that electric fields drop off via an inverse square law from a point charge. Gauss's law is particularly simplified in natural units:*

$$\nabla \cdot \vec{E} = \rho. \tag{7.1}$$

*You may have seen this with the permittivity of free space, ϵ_0, in the denominator on the right. However, in natural units, both ϵ_0 and the permeability of free space, μ_0, are equal to 1, because $1/\sqrt{\mu_0 \epsilon_0} = c$.

As a reminder, the divergence on the left can be expanded in Cartesian coordinates as

$$\nabla \cdot \vec{E} = \frac{\partial E^x}{\partial x} + \frac{\partial E^y}{\partial y} + \frac{\partial E^z}{\partial z}.$$

In other words, through a series of added first derivatives, it takes a vector and turns it into a scalar.

2. **Gauss's Law for Magnetism**

There's an almost identical law for magnetism, but with one crucial difference—the right-hand side is zero:

$$\nabla \cdot \vec{B} = 0. \tag{7.2}$$

There's a reason for this. There are no magnetic monopoles, thus you can't have a *clump* of magnetic charge.

Before we move on, though, it's worth observing that both of the first two equations contain first derivatives (in space, so far) of electric and magnetic fields and set them equal to something—charge density in the first case, zero in the second.

3. **Faraday's Law of Induction**

Faraday's law is kind of strange. A time-varying magnetic field can induce an electric field (which, in turn, induces electric currents). A conducting ring in a changing magnetic field, for instance, will generate a current that partially cancels the changing magnetic field.

There are a number of ways to express it, but the derivative form is quite instructive:

$$\nabla \times \vec{E} = -\frac{\partial \vec{B}}{\partial t}. \tag{7.3}$$

The term on the left, the **curl** of the E-field, is a little complicated to write out, but we'll do so for the sake of concreteness:

$$\nabla \times \vec{E} = \begin{vmatrix} \hat{i} & \hat{j} & \hat{k} \\ \partial_1 & \partial_2 & \partial_3 \\ E^1 & E^2 & E^3 \end{vmatrix} \tag{7.4}$$

$$= (\partial_2 E^3 - \partial_3 E^2)\hat{i} + (\partial_3 E_1 - \partial_1 E^3)\hat{j} + (\partial_1 E^2 - \partial_2 E^1)\hat{k}.$$

As Faraday's law is a vector equation, it is separable into components. The z-direction, for instance, may be written out as

$$\partial_1 E^2 - \partial_2 E^1 = -\partial_0 B^3.$$

4. **Ampère's Law**

Finally, we have Ampère's law—currents give rise to magnetic fields, as do changing electric fields:

$$\nabla \times \vec{B} = \frac{\partial \vec{E}}{\partial t} + \vec{J}, \tag{7.5}$$

again, greatly simplified from a more familiar form by writing it in natural units.

One thing you might notice about these laws is that the forms of the equations are all the same: they contain first derivates of E or B (or both) in space or time (or both), along with possible source terms, ρ or $\vec{J}$ (elements of our 4-current). But even if you've seen these equations before, what you may not yet realize is that, in relativity, *they are all the same equation*!

7.3   4-Potential

Electric and magnetic fields can dynamically give rise to one another. But even a decade before Maxwell, it was understood that both arise from various types of potential.

You are probably most accustomed to the idea of a gravitational potential energy in classical mechanics, typically labeled with $U(\vec{x})$ or $V(\vec{x})$. Potentials are defined throughout space and time, and while they are typically not particularly meaningful in their own right, derivatives give rise to forces:

$$\vec{F} = -\nabla U.$$

In Cartesian coordinates, the gradient can be expressed as

$$\nabla U = \frac{\partial U}{\partial x}\hat{i} + \frac{\partial U}{\partial y}\hat{j} + \frac{\partial U}{\partial z}\hat{k}. \tag{7.6}$$

In electromagnetism, the relationship between potential and field seems at first blush to be a bit more complicated. For one thing, there is a scalar potential, Φ, and a vector potential, $\vec{A}$. The electric and magnetic fields are related to them via the relationships

$$\vec{E} = -\nabla\Phi - \frac{\partial\vec{A}}{\partial t} \tag{7.7}$$

and

$$\vec{B} = \nabla \times \vec{A}. \tag{7.8}$$

As with gravity, the fields we care about and measure ($\vec{E}$ and $\vec{B}$) can be expressed as combinations of first derivatives of the scalar and vector potentials.

As we are now accustomed to thinking in 3+1 dimensional space, and we have a scalar potential, Φ, and a 3-vector, $\vec{A}$, why not combine them into a single object, the **4-potential**:

$$\vec{A}^{\mu} = \begin{pmatrix} \Phi \\ A^x \\ A^y \\ A^z \end{pmatrix}? \tag{7.9}$$

One of the exciting things about our relativistic intuition is that with a few clues (electromagnetic fields are first derivatives of potentials, all laws need to obey the tensor rules, source terms are related to first derivatives of electromagnetic fields) and a few good guesses, we can more or less write down all of electromagnetism from scratch.

7.4 The Faraday Tensor

First, we'll define a new object, the **Faraday tensor**:

$$F_{\mu\nu} \equiv \partial_{\mu}A_{\nu} - \partial_{\nu}A_{\mu}, \tag{7.10}$$

where A_{μ} is the lowered form of the 4-potential:

$$A_{\mu} = g_{\mu\nu}A^{\nu} = \begin{pmatrix} \Phi & -A^x & -A^y & -A^z \end{pmatrix}.$$

There's no deep reasoning behind lowering the indices on the Faraday tensor; we could just as easily compute $F^{\mu\nu}$ or F^{μ}_{ν}. They would be identical except for changes in minus signs. It just turns out that for ease of writing, the two lowered indices work nicely.

The Faraday tensor does have a few intriguing properties. For one thing, all of the terms are just linear combinations of first derivatives of potentials. We saw above that those are the properties of electromagnetic fields. The tensor is also antisymmetric (flop μ and ν and you get a minus sign), which means that the diagonal components are all zero (by definition) and the top six components are just the negatives of the bottom six. What are these six unique components?

Let's work them out explicitly. For instance, for $\mu = 0$ and $\nu = 1$,

$$F_{01} = -\partial_0 A_x - \partial_x \Phi$$

$$= -\dot{A}_x - \frac{\partial \Phi}{\partial x}$$

$$= E^x,$$

where the last step dropped out cleanly from comparison with equation 7.7. We need not go through every term in detail (though you're more than welcome to check), but likewise, $F_{02} = E^y$, $F_{03} = E^z$, $F_{10} = -E^x$, and so on.

What about the F_{12} component of the Faraday tensor?

$$F_{12} = \partial_x(-A^y) - \partial_y(-A^x)$$
$$= -(\nabla \times \vec{A})^z$$
$$= -B^z,$$

again, by comparison with equation 7.8. You can infer from this that $F_{23} = -B^x$ and $F_{31} = -B^y$. Combining all of these components, we can write out the Faraday tensor explicitly:

$$F_{\mu\nu} = \begin{pmatrix} 0 & -E^x & -E^y & -E^z \\ E^x & 0 & B^z & -B^y \\ E^y & -B^z & 0 & B^x \\ E^z & B^y & -B^x & 0 \end{pmatrix}, \tag{7.11}$$

where the first index denotes the column and the second denotes the row.

We can also raise both indices (which will be useful for other purposes). As a practical matter, the two are identical up to a switch in sign for the electric fields:

$$F^{\mu\nu} = \begin{pmatrix} 0 & E^x & E^y & E^z \\ -E^x & 0 & B^z & -B^y \\ -E^y & -B^z & 0 & B^x \\ -E^z & B^y & -B^x & 0 \end{pmatrix}. \tag{7.12}$$

The Faraday tensor represents a general way of writing down electromagnetic fields in terms of the 4-potential without ever having to specify which is which.

Example: Boosting an E-Field

As an example of how this plays out, let's go back to the situation in figure 7.1 and 7.2, in which we have a charged, infinite wire that is boosted along the direction of the wire. We're looking at a point some point above (in the y-direction) a wire, with an E-field (only) in the E^y direction:

$$F^{\mu\nu} = \begin{pmatrix} 0 & 0 & E_0 & 0 \\ 0 & 0 & 0 & 0 \\ -E_0 & 0 & 0 & 0 \\ 0 & 0 & 0 & 0 \end{pmatrix},$$

with all other components equal to zero. Then, we boost the whole frame to the positive x-direction:

$$\Lambda^{\bar{\mu}}_{\mu} = \begin{pmatrix} \gamma & v\gamma & 0 & 0 \\ v\gamma & \gamma & 0 & 0 \\ 0 & 0 & 1 & 0 \\ 0 & 0 & 0 & 1 \end{pmatrix}.$$

But there's a trick. We need to apply this transformation *twice*:

$$F^{\bar{\mu}\bar{\nu}} = \Lambda^{\bar{\mu}}_{\mu} \Lambda^{\bar{\nu}}_{\nu} F^{\mu\nu}$$

This calculation seems rough, except for the fact that we only have a couple non-zero terms.

Multiplying it out,

$$\begin{pmatrix} \gamma & v\gamma & 0 & 0 \\ v\gamma & \gamma & 0 & 0 \\ 0 & 0 & 1 & 0 \\ 0 & 0 & 0 & 1 \end{pmatrix} \begin{pmatrix} 0 & 0 & E_0 & 0 \\ 0 & 0 & 0 & 0 \\ -E_0 & 0 & 0 & 0 \\ 0 & 0 & 0 & 0 \end{pmatrix}$$

$$\begin{pmatrix} \gamma & v\gamma & 0 & 0 \\ v\gamma & \gamma & 0 & 0 \\ 0 & 0 & 1 & 0 \\ 0 & 0 & 0 & 1 \end{pmatrix} = \begin{pmatrix} 0 & 0 & \gamma E_0 & 0 \\ 0 & 0 & v\gamma E_0 & 0 \\ -\gamma E_0 & -v\gamma E_0 & 0 & 0 \\ 0 & 0 & 0 & 0 \end{pmatrix}.$$

In the moving frame, the E-field seems stronger by a factor of γ. Why? Because the charge density seems higher.

And there's a B-field in the +z-direction (compare this to the general form in equation 7.11. If you wish to check your intuition, you can check in with the right-hand rule (or figure 7.2). Generate a current to the right in the wire, and you get a +z B-field.

7.5 Maxwell's Laws in Special Relativity

So that's all the background you need—though admittedly, it's a lot. Now, let's make a more elegant version of *all* of electromagnetism. (Note that by "more elegant," I don't mean that the math won't appear just a little bit ugly.)

7.5.1 THE BIANCHI IDENTITY

We can do a number of interesting tricks based on only the structure of the Faraday tensor itself. As Juliet [40] notes:

> That which we call a rose
> By any other name would smell as sweet.

More prosaically, the labels on our variables and indices are completely arbitrary. For instance, consider the set of first derivatives of the Faraday tensor (eq. 7.10).
Order 1:

$$\partial_\lambda F_{\mu\nu} = \partial_\lambda \partial_\mu A_\nu - \partial_\lambda \partial_\nu A_\mu.$$

This is a rank-3 tensor, in principle—a collection of 64 numbers, many of them redundant. For instance, it doesn't matter in what order you take derivatives. So $\partial_\lambda \partial_\mu A_\nu = \partial_\mu \partial_\lambda A_\nu$. But even so, we're not going to work out these terms explicitly.

But heeding Juliet's advice, there's no particular reason to label the first index λ, the second, μ and the third ν. We could sort them however we like.
Order 2:

$$\partial_\nu F_{\lambda\mu} = \partial_\nu \partial_\lambda A_\mu - \partial_\nu \partial_\mu A_\lambda.$$

Order 3:

$$\partial_\mu F_{\nu\lambda} = \partial_\mu \partial_\nu A_\lambda - \partial_\mu \partial_\lambda A_\nu.$$

We're *allowed* to reorder the indices, but just because we *can* doesn't mean that we *should*. But we have a trick up our sleeves. Add the three orders together and something interesting happens:

$$\partial_\lambda F_{\mu\nu} + \partial_\mu F_{\nu\lambda} + \partial_\nu F_{\lambda\mu} = \partial_\lambda \partial_\mu A_\nu - \partial_\lambda \partial_\nu A_\mu + \partial_\nu \partial_\lambda A_\mu$$

$$- \partial_\nu \partial_\mu A_\lambda + \partial_\mu \partial_\nu A_\lambda - \partial_\mu \partial_\lambda A_\nu$$

$$= 0. \tag{7.13}$$

It may take you a while to assure yourself that every term on the right identically cancels, but cancel they do.[*] The resulting expression is known as the **Bianchi identity**. Though the relation represents 64 possible identities (four values each for μ, ν, and λ), in reality, it's far fewer, since permutations of the three indices produce identical relations. Further, there are lots of trivial results, as is the case when any two indices repeat. The only interesting identities are those for which all three indices are different.

For example, it we expand the Bianchi identity for $\mu = 0$, $\nu = 1$, and $\lambda = 2$, we get

$$\partial_y E^x + \dot{B}^z + \partial_x(-E^y) = 0.$$

Or, equivalently,

$$(\nabla \times \vec{E})^z = -\dot{B}^z.$$

Combining all of the nonvanishing Bianchi identity terms yields

$$\boxed{\nabla \times \vec{E} = -\dot{\vec{B}}} \text{—Faraday's law of inductance.} \qquad (7.14)$$

Likewise, if we expand the Bianchi identity for the case in which $\mu = 1$, $\nu = 2$ and $\lambda = 3$, we get

$$\partial_z B^z + \partial_x B^x + \partial_y B^y = 0,$$

or, in the more familiar form,

$$\boxed{\nabla \cdot \vec{B} = 0} \text{—Gauss's law for magnetism.} \qquad (7.15)$$

The *form* of the Faraday tensor immediately generated two of Maxwell's equations.

7.5.2 ONE EQUATION!

I introduced the idea of electromagnetism in relativity because we knew that whatever we got, it would have to be Lorentz invariant. But we also saw, just by looking at Maxwell's equations, that the form needs to be something like

$$\text{deriv. of } F^{\mu\nu} = \text{source.}$$

[*] First and sixth, second and third, fourth and fifth.

There honestly aren't that many ways you can get such an equation and still have it obey the tensor rules. There is (except for multiplication by a constant) really only one:

$$\boxed{\partial_\mu F^{\mu\nu} = J^\nu}, \tag{7.16}$$

which we can multiply out for $\nu = 0$:

$$\partial_\mu F^{\mu 0} = \frac{\partial F^{00}}{\partial t} + \frac{\partial F^{10}}{\partial x} + \frac{\partial F^{20}}{\partial y} + \frac{\partial F^{30}}{\partial z}$$

$$= \frac{\partial E^x}{\partial x} + \frac{\partial E^y}{\partial y} + \frac{\partial E^z}{\partial z}$$

$$= \nabla \cdot \vec{E},$$

where $J^0 = \rho$. Putting it all together, we get

$$\boxed{\nabla \cdot \vec{E} = \rho} \quad \text{—Gauss's Law.} \tag{7.17}$$

Now, let's try $\nu = 3$:

$$\partial_\mu F^{\mu 3} = \frac{\partial F^{03}}{\partial t} + \frac{\partial F^{13}}{\partial x} + \frac{\partial F^{23}}{\partial y} + \frac{\partial F^{33}}{\partial z}$$

$$= \frac{\partial(-E^z)}{\partial t} + \frac{\partial(B^y)}{\partial x} + \frac{\partial(-B^x)}{\partial y}$$

$$= -\frac{\partial E^z}{\partial t} + (\nabla \times \vec{B})^z$$

$$= J^z$$

$$(\nabla \times \vec{B})^z = \frac{\partial E^z}{\partial t} + J^z.$$

We'd get a similar result for $\nu = 1, 2$, which can be combined to yield

$$\boxed{\nabla \times \vec{B} = \dot{\vec{E}} + \vec{J}} \quad \text{—Ampère's law.} \tag{7.18}$$

Pretty amazing!

There is one more exciting wrinkle. Let's take another derivative of our generalized Maxwell equation (eq. 7.16):

$$\text{(A)} \ \partial_\nu \partial_\mu F^{\mu\nu}.$$

We can do this term by term, if we like, but consider swapping the order:

$$(B) \; \partial_\mu \partial_\nu F^{\nu\mu}.$$

But here's the thing: (A) and (B) are identical. Because the order of derivatives is irrelevant, adding (A) and (B) can be written as

$$\partial_\nu \partial_\mu \left(F^{\mu\nu} + F^{\nu\mu} \right) = 0,$$

because the Faraday tensor is antisymmetric. But that means both (A) and $(B) = 0$, or

$$\partial_\nu \partial_\mu F^{\mu\nu} = \partial_\nu J^\nu = 0.$$

In other words, we've actually just *proven* that charge is conserved.

7.5.3 RADIATION

The relativistic version of Maxwell's equations (eq. 7.16) can also shed some light on light, especially once we take sources out of the equation. Doing so quickly yields

$$\partial_\mu F^{\mu\nu} = \partial_\mu \partial^\mu A^\nu - \partial_\mu \partial^\nu A^\mu. \tag{7.19}$$

This already looks pretty compact, but we can make it simpler still by noting that, in terms of electromagnetism, the fundamental "objects" aren't the potential fields. As with ordinary gravitational potential, we could add a constant to the 4-potential, and as the derivatives vanish, this addition would produce no additional force on electrical charges—and thus wouldn't be detectable.

But we can do more than simply add a constant. Let's suppose that we added a contribution to the field using the form

$$A^\mu \rightarrow A^\mu + \partial^\mu \theta(x).$$

In principle, $\theta(x)$ is completely arbitrary. It could be a constant, or a gradient, a big lump that quickly disappears. It could vary in time or space or not at all. This is known as a **gauge freedom**, and one of the nice things about gauge freedoms is that we never actually need to produce the gauge field, $\theta(x)$; we just need to know that we could if we want to.

What's more, regardless of what we pick, the additional potential will contribute nothing to the Faraday tensor:

$$dF^{\mu\nu} = \partial^\mu \partial^\nu \theta - \partial^\nu \partial^\mu \theta = 0.$$

Knowing that we're allowed to change the 4-potential in this way is incredibly useful. For example, suppose we wanted to make the 4-divergence of the potential vanish:

$$\partial_\mu A^\mu = 0.$$

This isn't automatic. Imagine that the divergence is nonzero and you wanted to get rid of it. We could apply a transformation of the form:

$$\partial_\mu A^{\mu(new)} = \partial_\mu A^{\mu(old)} + \partial_\mu \partial^\mu \theta = 0.$$

It would be a giant pain to actually compute the field, θ that would make this work. But fortunately, we don't need to. We simply assert that such a field exists, and thus, the 4-divergence vanishes.

If this condition is met (so-called **Lorenz gauge**[*]), then equation 7.19 reduces to

$$\partial^\mu \partial_\mu A^\nu = 0. \tag{7.20}$$

We went to all of this effort for a couple of reasons. For one, it means that each component of the 4-potential wave seems to be evolved independently from one another, and secondly, because the operator, $\partial_\mu \partial^\mu$, is extremely important in wave mechanics. It is known as the **d'Alembertian**, and is sometimes written as

$$\partial_\mu \partial^\mu \mathbf{A} = \Box \mathbf{A}.$$

If we expand it term by term, it takes the form

$$\ddot{\mathbf{A}} - \nabla^2 \mathbf{A} = 0, \tag{7.21}$$

where ∇^2 is the **Laplacian**, and can also be expanded out in Cartesian coordinates as

$$\nabla^2 \Phi = \frac{\partial^2 \Phi}{\partial x^2} + \frac{\partial^2 \Phi}{\partial y^2} + \frac{\partial^2 \Phi}{\partial z^2}, \tag{7.22}$$

which often shows up in the context of the Poisson equation in gravitation and electrostatics.

The combination of double time derivative and Laplacian is one of the most common and straightforward second-order differential equations. Equation 7.21 is solved as

$$A^\nu \propto \exp(\vec{k} \cdot \vec{x} - \omega t),$$

where $\omega/|\vec{k}| = 1$. In other words, it predicts a wave of potential propagating through space at the speed of light. This is, in fact, a beam of light.

---

[*] That's not a typo, though you'll often seen it misattributed to Lorentz, especially given how much else his name is attached to. Lorenz and Lorentz are two different people.

7.6   The Lorentz Force Law

We've described how electric charges can create electromagnetic fields, but we also need to figure out how electromagnetic fields can apply forces on charged particles. This is known as the **Lorentz force law**, and you may know it as

$$\vec{F} = q(\vec{E} + \vec{v} \times \vec{B}). \tag{7.23}$$

You can get a sense of a particle interacting in an electromagnetic field in figure 7.4.

As with Maxwell's equations, we can *guess* at what the relativistic version of the Lorentz law might look like, starting with

$$(\text{Force}) = q \times F^{\mu\nu} \times (\text{v } or \text{ 1}).$$

While we'd normally write the force part of the equation as

$$(\text{Force}) = \frac{dp^i}{dt},$$

there are two complications: (1) All equations should be written exclusively in terms of generalized 3+1 dimensional spacetime coordinates and (2) we can't have derivatives with respect to t, which would make the equation specific to one particular frame. Instead, let's make the left-hand side

$$\frac{dp^\mu}{d\tau}.$$

As for the right-hand side, we can't simply have $qF^{\mu\nu}$. Why? Because the indices on the right wouldn't match the left. This can be fixed if we contract over U_ν.

Putting it together, our guess is

$$\frac{dp^\mu}{d\tau} = qU_\nu F^{\mu\nu}. \tag{7.24}$$

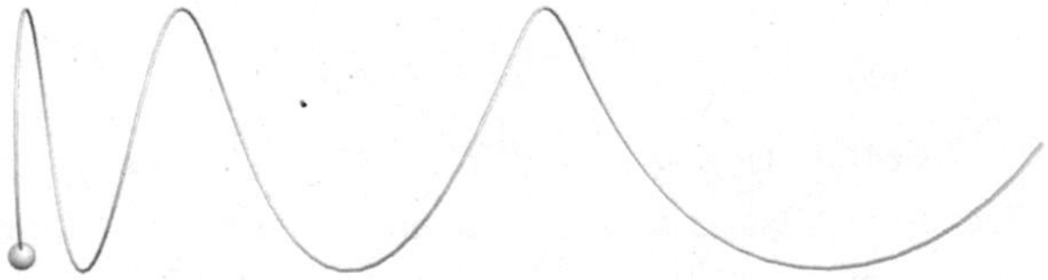

FIGURE 7.4. An electron in an E- and B-field. Both fields are pointing to the right. The electron moves in a helix (due to the B-field), and increases in speed (due to the E-field).

It turns out that this guess is exactly right! For $\mu = 1$, for instance,

$$\frac{dp^x}{d\tau} = qU_\mu F^{\mu 1}$$
$$= qU_0 F^{10} + qU_2 F^{12} + qU_3 F^{13}$$
$$= q\gamma \left[E^x - v^y B^z + v^z B^y \right]$$
$$\frac{dp^x}{dt} = q\left[\vec{E} + \vec{v} \times \vec{B} \right]^x.$$

Exactly as predicted!

While we wrote down the Lorentz-invariant form of the force law with the classic, spatial version in mind, there is an added bonus. Using $\mu = 0$, we get a power relation:

$$\frac{dE}{dt} = q\vec{E} \cdot \vec{v}.$$

The power applied by an electromagnetic field is a consequence of only the electric field—not the magnetic part!

Looking Forward

Congratulations! You can now legitimately claim to *know* special relativity, including much of the major findings from Einstein, Minkowski, and others in the aftermath of the miracle year. However, we're not quite done. Through the example of electromagnetism, we can get an idea of how other force laws might be developed, including the force of gravity. In the final chapter, we'll take a big step toward thinking about how gravity might work in a relativistic universe.

7.7 Problems

1. In Section 7.1, we introduced two parallel wires with charge density λ in the boosted frame. For two infinite wires, the electric field is

$$E_r = \frac{\lambda}{4\pi r},$$

in natural units. Thus, the force per unit length is

$$\frac{F_r}{L} = \frac{\lambda^2}{2\pi r}.$$

Likewise, the magnetic field for a current-carrying wire is

$$B_\perp = \frac{I}{2\pi r},$$

with the magnetic force per unit length just equal to the product of the current times the magnetic field.

(a) What is the electrical force per unit length between the two wires in the boosted frame? Answer in terms of λ, r, v, and γ. (Don't forget the lessons of the previous chapter that charge density increases in a boosted frame.)

(b) What is the magnetic force per unit length (attractive) between the two wires in the boosted frame?

(c) How does the net force per unit length compare to the net force per unit length in the unboosted frame?

(d) What are the implications for the net accelerations between the two wires?

2. A short problem to flex your classical electromagnetism muscles: imagine that you have a sheet of material extending infinitely in the y- and z-directions), and from $x = -L/2$ to $L/2$. It has charge density distribution that only varies in the x-direction:

$$\rho(x) = \rho_0 \frac{x}{L}.$$

What is the electric field in the interior of the sheet?

Hint: The sheet is electrically neutral overall, so outside the sheet, the electric field vanishes.

3. In the example in figures 7.1 and 7.2, we consider a "pure" electric field in the y-direction. We then boost the system into the positive x-direction by a velocity v. Let's look at a very similar problem: Start with a pure electric field, E_0, in the positive z-direction (no B-fields).

(a) Write down $F^{\mu\nu}$ for this system.

(b) Boost this field by a speed v in the positive x-direction. Be sure to follow the example. Write down the boosted Faraday tensor, $F^{\overline{\mu}\overline{\nu}}$.

(c) In the boosted frame, write down the E-field and B-field in terms of E_0.

In case you're unclear about the goal of this problem, the example in the text would result in $\vec{B} = E_0 \gamma \hat{k}$.

4. Consider a 4-potential of the form

$$A^\mu = \begin{pmatrix} 0 \\ 0 \\ B\sin(kz - \omega t) \\ 0 \end{pmatrix}.$$

(a) Calculate $\partial_\mu A^\mu$.

(b) Calculate all nonzero elements of $F_{\mu\nu}$. What are the nonzero elements of $\vec{E}$ and $\vec{B}$ fields?

(c) Calculate $\partial_\mu F^{\mu\nu}$ under the assumption of $\omega = k$. What does this imply about the speed of propagation and the local electric charge density?

5. Consider the Lorentz-invariant form of the Lorentz force law (eq. 7.24) with an electron moving in the positive x-direction at speed v and corresponding gamma-factor γ.

   (a) Write down the 4-velocity of the electron, U_ν, in 1-form notation (downstairs).

   (b) Now consider an electromagnetic field where the E-field has a strength of E_0 and is pointed in the x-direction and the B-field has a strength of B_0 and is pointed in the y-direction.

   Write down the Faraday tensor, $F^{\mu\nu}$, for this case.

   (c) Using equation (7.24), compute $dp^\mu/d\tau$ for the electron.

   (d) What is the power dp^0/dt (note the denominator is dt, not $d\tau$) on the electron?

FURTHER READINGS

- Albert, Einstein, "On the Electrodynamics of Moving Bodies," Annalen der Physik 17, no. 10 (1905): 891–921. (Translation by Meghnad Saha, 1920.) Normally I reserve technical references for the bibliography at the end of the book. However, this paper is particularly readable, and some students may find both the completeness of the theory (Einstein includes a fairly modern set of the Lorentz transforms) and notational differences of interest.

- Richard Feynman, QED: The Strange Theory of Light and Matter (Princeton University Press, 2014). While this chapter has focused on turning classical electromagnetism into a relativistic theory, Feynman's work was instrumental to quantizing the theory. QED was written at an introductory level that should be very accessible to you at this point, and which will give a great idea as to the next steps in physical sophistication.

- David Griffiths, Introduction to Electrodynamics, 4th ed. (Cambridge: Cambridge University Press, 2017). Griffiths has written several gold-standard undergraduate physics texts. For those looking for a review, or who want to learn on your own about how classical electrodynamics works, this is a very good introduction.

A Brief Introduction to General Relativity

Albert Einstein (1879–1955) circa 1905. Einstein developed many of the principles of special relativity but was arguably even more central to general relativity. By 1907, he developed much of the reasoning that we'll discuss in this chapter.

8.1 The Equivalence Principle

What makes special relativity so special is that, axiomatically, there's no gravity. We've seen, however, that we can simulate gravity by simply hitting the gas. Drive in a car or ride in an elevator, and you've experienced it for yourself. Einstein was thinking along precisely these lines in 1907 [11], when he first started work on his theory of gravity, general relativity. His motivation was encapsulated by the **equivalence principle**, in which he invites to the reader to

... assume the complete physical equivalence of a gravitational field and a corresponding acceleration of the reference system.

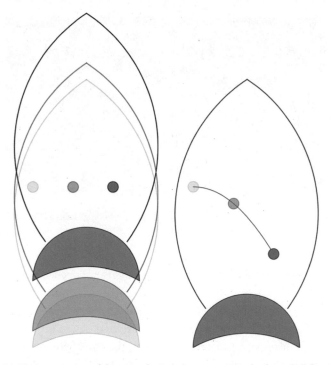

FIGURE 8.1. Two perspectives of the same physical phenomenon. In the first, a ball flies to the right past the window of an accelerating spaceship. The ball is not accelerating at all, but the spaceship is. Seen from the window of the spaceship (right frame), the ball appears to be "falling" toward the floor of the ship along a parabolic arc, exactly as though it were in a gravitational field.

That is, there's no functional, observational, or measurable distinction between a *real* gravitational field and "artificial" gravity generated by accelerating your car or hitting the thrusters on your rocket, except for the view out of your windows or the rumbling of the ship.

We explored the mathematics of an accelerated frame back in Chapter 3. But there's another subtle implication. Consider figure 8.1, in which a projectile is viewed in an accelerated frame. In this case, the ball appears to fall toward the back of a spaceship.

The equivalence principle tells us that all bodies (in the absence of air resistance) fall with the same acceleration, as predicted by Galileo. In an accelerated box, the floor rushes up to meet a feather and a hammer at the same rate, which means that the same holds true in a gravitational field. Indeed, this exact experiment was conducted during the Apollo 15 mission in 1971 (figure 8.2).

Another implication of the equivalence principle is that an object will accelerate toward the floor regardless of mass. A projectile could be massless—a stream of photons—and it would still arc downward.

FIGURE 8.2. In 1971, Apollo 15 astronaut David Scott dropped a feather and a hammer on the Moon and found that, in the absence of air resistance, they dropped at the same rate.

Before we delve further into gravitational fields, we need to take baby steps to deal with the apparent *absence* of apparent gravitational fields. No one would deny that there is gravity near the surface of the Earth,* but we can imagine and even implement an experiment to make even the meager pull of Earth seem to disappear. Einstein's corollary, the so-called *weak* equivalence principle, posits that the physics of free fall are identical to the physics of no gravity—or, in other words, the world of special relativity.

Consider NASA's "Vomit Comet," a plane that climbs about 5 miles into the atmosphere in a steep upward trajectory and then essentially cuts the gas, leaving the plane in free fall. From inside, the passengers feel as though they are weightless, even though the plane itself is very much still inside Earth's gravitational field. Why? Because even as the passengers are in free fall, the plane is as well. And since the passengers aren't accelerated relative to the Vomit Comet, they appear to be in an inertial frame.

There is one wrinkle to the weak equivalence principle. You can imagine, for instance, that if I put you in a big elevator and drop you toward the Earth you won't be able to tell that you're falling, no matter what you do in your freely falling frame. You'll float around, and everything will appear to be in 0 g. But if the elevator is very, very big, someone inside could potentially measure gravitational differentials through tidal forces.

So although in principle you could tell whether you're in deep space or being dropped by making very precise measurements at the top and bottom of your box, as a practical matter *locally*, we can always have a freely falling

---

*Though we'll see, soon enough, in our geometrized units, that Earth's gravity is incredibly weak.

observer. And that's the realm where special relativity is at play. That is, indeed, what puts the "special" in special relativity.

8.2   Rotating Reference Frames

Rather than focus on the clichéd artificial gravity of an accelerating rocket ship, let's return to a rigidly rotating reference frame last seen in our discussion of the Ehrenfest paradox. To make matters concrete, imagine a world of superintelligent ants living atop a turntable, spinning with angular frequency ω (figure 8.3). The world is, for all intents and purposes, two-dimensional, and the ants stay in place due to their friction with the floor. Most important, there is no true gravitational field in this scenario.

The ants near the center are barely moving. Though they spin, we'll assume the rate is gentle enough that they only feel a very slight bulging outward on all their sides. But many ants live farther away from the center, and the farther out they are, the faster they move:

$$v = r\omega,$$

since this universe consists of a rigid rotator. The ants experience a centrifugal force, which feels to them to be an artificial gravity of

$$g = \frac{v^2}{r} = r\omega^2.$$

The ants know nothing of their rotating world. The only feel an outward tug. Moreover, the farther from the center they are, the stronger the pull. This is very much like an actual hill, where the center is the peak. The ant physicists can even compute gravitational potential:

$$\Phi_G = -\int g \, dr = -\frac{1}{2}r^2\omega^2. \tag{8.1}$$

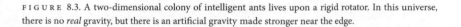

FIGURE 8.3. A two-dimensional colony of intelligent ants lives upon a rigid rotator. In this universe, there is no *real* gravity, but there is an artificial gravity made stronger near the edge.

Consider an ant at radius r. They think they're standing still in a gravitational field, but we (omniscient three-dimensional creatures that we are) know that in truth they're moving at a speed directly proportional to their distance from center. At higher and higher speeds, they experience greater and greater time dilation:

$$\gamma = \frac{1}{\sqrt{1 - v^2}}$$

$$\approx 1 + \frac{v^2}{2}$$

$$= 1 + \frac{r^2\omega^2}{2}$$

$$\frac{t}{t_0} = 1 - \Phi_G.$$

Aha! Remember that Φ is negative, so time runs slower the farther an ant is from the center. The central conjecture of the equivalence principle is that there is no distinction between accelerated reference frames and real gravitational fields. Thus, if the clocks run slower in the *down* direction for ants, then it should also run slower in the down direction on Earth.

So the potential gravity on Earth

$$\Phi_G = -\frac{GM}{R} = 6.25 \times 10^7 \text{ m}^2/\text{s}^2 = 6.9 \times 10^{-10}.$$

where I've converted to natural units. When we talk about gravity on Earth being "weak," this is what we mean. The dimensionless potential is many, many times smaller than 1.

8.3   Gravitational Redshifting

Without actually deriving the Einstein field equations, the curvature of spacetime or any of the nuts and bolts of general relativity, our goal throughout this chapter is to try to exploit the tools of special relativity to make some predictions about what general relativity might look like. Contrary to our previous examples, our next example explicitly invokes Newtonian gravity and energy conservation in order to induce us to make an educated guess of how to resolve the two.

For the purpose of this thought experiment, let's ignore thermal energy and simply pretend that all forms of energy are infinitely fungible. Consider a neutral, ground state hydrogen atom of mass m. Creating such an atom "from

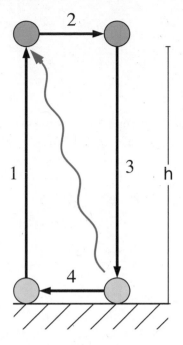

FIGURE 8.4. An atom goes through a continuous cycle of photon emission and potential energy change, and, in the process, apparently creates perpetual motion.

scratch," as it were, requires $E_M = mc^2$ of energy. We then lift the atom up a high cliff, h, as shown in Step 1 of figure 8.4. The requires a total energy of

$$E_1 = mc^2 + mgh,$$

where, for the time being, we're leaving in factors of c^2 explicitly.

At this point, we excite the atom with a photon, E_γ. For the first excited state of hydrogen, this would simply be 10.2 eV, but as we've seen, the excited state of hydrogen will have a mass of

$$m^* = m + \frac{E_\gamma}{c^2}.$$

The excited atom is dropped off the cliff (Step 3), and the potential energy, $m^* gh$, is liberated. At the bottom of the cliff, the atom is allowed to de-excite, sending off a photon of energy E_γ. We are left with what we started, a ground state atom at the bottom of the cliff. However, when we look over the whole cycle, it seems like we've mysteriously conjured energy from thin air:

$$\Delta E = \left(m^* gh + E_\gamma - mc^2 \right) - \left(mc^2 + mgh + E_\gamma \right)$$
$$= (m^* - m)gh$$
$$= E_\gamma gh.$$

What?

The only way around this conundrum is to assume that we've miscalculated somehow, that the photon loses an equivalent amount of energy on the way up the cliff. That is,

$$\frac{\Delta E}{E_\gamma} = -\frac{gh}{c^2},$$

or, more generally,

$$\frac{\Delta E}{E_\gamma} \approx -\frac{\Delta \Phi_G}{c^2}, \qquad (8.2)$$

for an arbitrary gravitational field.

Light loses energy as it gets fired up out of a gravitational field. This is known as **gravitational redshifting.** This redshift is regularly observed from spectral lines in white dwarves, but it can even be measured on Earth. In 1959, Robert Pound and Glen Rebka performed an experiment at Harvard's Jefferson Laboratory in which gamma rays were sent up a height of only 22.5 m, but the effect was still measurable [36].

We can even (mostly) predict the existence of black holes from equation 8.2. If we envision a point mass, M, at the origin, then starting from a distance of r, the potential required to escape to infinity is GM/r. Thus, for a photon to lose all of its energy, we might suppose (wrongly) that if $GM/rc^2 = 1$, the photon will escape. This is solved by

$$r_{BH} \overset{?}{=} \frac{GM}{c^2}.$$

Close, but no cigar. Equation 8.2 was only approximate, and in actual fact, the radius of a black hole—the limit from which light can't escape—is

$$R_{BH} = \frac{2GM}{c^2}. \qquad (8.3)$$

8.4   What General Relativity Might (and Does) Look Like

We spent a considerable amount of time in the previous chapter thinking about how the Lorentz-invariant demands of relativity influence the tensor forms of Maxwell's and the Lorentz force equations, and considerable time in this chapter thinking about the observational and experimental predictions of general relativity by simply extrapolating from special relativity. We'll close by merging the two and thinking about what the form of general relativity might look like.

Newtonian gravity is governed by two key equations. In Cartesian coordinates, the acceleration equation may be written as

$$\ddot{x}^i = \partial^i \Phi_G, \tag{8.4}$$

which looks pretty good from a tensorial point of view. But it's not great. For one thing, the derivatives are only taken over three spatial dimensions. For another, the left-hand side has a time derivative, which isn't Lorentz invariant.

But it's actually worse. The gravitational potential is dependent on the frame of reference, or should be. The second governing equation in Newtonian gravity, the Poisson equation, is

$$\nabla^2 \Phi_G = -\partial_i \partial^i \Phi_G$$
$$= 4\pi G \rho, \tag{8.5}$$

which, again, looks consistent with tensor rules in three-dimensional space but not in spacetime. There are several problems with equation 8.5 that arise from trying to reconcile it with special relativity.

For one, the potential field seems to be generated instantaneously by changes in the density, ρ. That is, the gravitational effects of a moving star would be *seen* across the galaxy immediately. Based on everything we've discussed so far, a good theory of gravity *should* propagate at the speed of light (or slower).

Another issue is that the source term for gravity shouldn't just include mass. As we've already seen, energy can change forms to and from mass relatively easily, and a conversion of form shouldn't result in a change of gravity.

Finally, we've already seen that even the energy density isn't sufficient to describe the energy in a Lorentz-invariant way. Rather, we require the stress-energy tensor, $T^{\mu\nu}$.

If we were to guess, we'd suppose that general relativity should look *something* like

$$G^{\mu\nu} = 8\pi T^{\mu\nu}, \tag{8.6}$$

where I've included the 8π normalization factor (purely a convention) because doing so doesn't change the argument at all. $G^{\mu\nu}$ (known as the Einstein tensor) is a bunch of second derivatives over space, and, presumably, time.

Looking back at equation 6.10, if the flow of matter and the sound speed in fluids are slow compared to the speed of light—the very conditions that govern nonrelativistic gravity—then the stress-energy tensor is simply

$$T^{\mu\nu}(NR) = \begin{pmatrix} \rho(x) & 0 & 0 & 0 \\ 0 & 0 & 0 & 0 \\ 0 & 0 & 0 & 0 \\ 0 & 0 & 0 & 0 \end{pmatrix}.$$

We might suppose, then, that to make the Einstein tensor align with Newtonian gravity, it could be

$$G^{\mu\nu}(NR) \overset{?}{=} \begin{pmatrix} 2\nabla^2\Phi_G & 0 & 0 & 0 \\ 0 & 0 & 0 & 0 \\ 0 & 0 & 0 & 0 \\ 0 & 0 & 0 & 0 \end{pmatrix}.$$

What's more, whatever the Einstein tensor is a second derivative of (it's the metric, in case you are curious), *forces* are going to be generated by first derivatives of the same.

We still have a problem, however. Suppose that the gravitational sources aren't stationary. Even if all we're considering is the ordinary scalar potential energy, we still haven't accounted for the fact that a gravitational signal seems to propagate at the speed of light. A much better guess would be of the form

$$G^{\mu\nu}(NR) = \begin{pmatrix} -2\square\Phi_G & 0 & 0 & 0 \\ 0 & 0 & 0 & 0 \\ 0 & 0 & 0 & 0 \\ 0 & 0 & 0 & 0 \end{pmatrix}.$$

As with light, this means that gravitational signals—gravitational *waves*—propagate at the speed of light. It is, after all, the only mathematical structure that yields Lorentz invariance. It means, for instance, that in a pair of orbiting white dwarves, the change in potential won't be seen instantaneously but, rather, the retarded potential will be the one measured by a distant accelerometer.

All of this guesswork arises from a specific weak-field solution. It works on Earth, but not near black holes. If you'd like to learn more about what happens in the strong gravity limit, I've got some excellent news. Having mastered flat spacetime, you're fully prepared to enter the world of curvature.

8.5   Problems

1. To get a sense of the distinction between strong and weak gravitational fields, calculate the dimensionless potential

$$\Phi_G = -\frac{GM}{rc^2}$$

for the following systems. You may need to look up a few physical quantities.
   (a) The surface of the Earth
   (b) The surface of the Sun
   (c) The surface of a white dwarf (For concreteness, use $M = 2 \times 10^{30}$ kg and $R = 7 \times 10^6$ m.)
   (d) The surface of a neutron star (Use $M = 2.8 \times 10^{30}$ kg and $R = 1.1 \times 10^4$ m.)

2. In the Pound-Rebka experiment, gamma rays are emitted on the ground and reabsorbed at a height of 22.5 m. What fraction of the photon energy is lost?

3. As an alternative proof of concept regarding gravitational redshift as well as gravitational lensing, consider a photon shot through an instantaneously stationary spaceship as shown (time progression goes from fainter to darker). The spaceship is just starting its acceleration at a rate g.

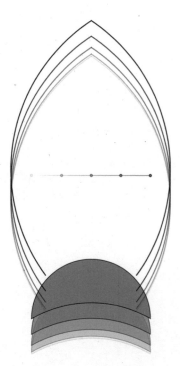

   Solve all parts in terms of g, L, the width of the spaceship, and ν_0, the emitted frequency of the photon. Include factors of c explicitly.

(a) As seen from outside the spaceship, how long does it take the photon to cross the ship?

(b) By that time, assuming the spaceship is still nonrelativistic, how far will it have moved? This, presumably, is how far the photon will appear to be bent toward the bottom of the spaceship.

(c) At that instant, what is the speed of the spaceship?

(d) To leading order, what is $\gamma - 1$ (e.g., the relative time dilation factor as well as the blueshift fraction, $\Delta\nu/\nu_0$)?

(e) If the potential (as seen within the spaceship) is $\Delta\Phi_G = g\Delta y$, what is the apparent blueshift in terms of the potential?

FURTHER READINGS

- Sean M. Carroll, *Spacetime and Geometry* (Cambridge University Press, 2019). An excellent and fairly comprehensive discussion of introductory and intermediate gravitational physics using similar notation and conventions as the current book.

- Richard Feynman, Matthew Sands, and Robert B. Leighton, *The Feynman Lectures on Physics* (Massachusetts: Addison-Wesley, 1963). The entire series of lectures is absolutely worth a read, but especially Lecture 42 on "Curved Space." My discussion of time near the top and bottom of an accelerated frame borrows heavily from Feynman's discussion.

- Charles W. Misner, Kip S. Thorne, and John A. Wheeler, *Gravitation* (Princeton University Press, 2017). This book is a classic of general relativity, and most physicists obtain a copy for reference. It is very well written and has the answer to just about every relativistic question.

- Bernard Schutz, *A First Course in General Relativity*, 3rd ed. (Cambridge University Press, 2022). I have mentioned this book earlier, as it provides the basis for much of our notation. Our work in the current volume leads very smoothly into the level of rigor and difficulty in Schutz's book.

Appendix A

Notation and Tensor Rules

A.1 Notation

Throughout the text, we've endeavored to use consistent notation:

Symbol	Description	First Used
x^i or u^i or v^i	Italicized, with a Roman index (e.g., position, particle velocity, or frame velocity). The components of a 3-vector: (1, 2, 3).	p. 4
$\vec{x}$, $\vec{u}$, or $\vec{v}$	The 3-vector (e.g., position, particle, or frame velocity).	p. 4
u_i or U_μ	Italicized, with a lowered index. The components of a 1-form (Roman index for space, Greek for spacetime).	p. 8
g_{ij} or $g_{\mu\nu}$	The 2-form version of the metric tensor. Our convention is $g_{00} = 1$.	p. 9
$x^{\bar{i}}$ or $x^{\bar{\mu}}$ or $\left(\bar{t}, \bar{x}, \bar{y}, \bar{z}\right)$	The components of a vector in a barred coordinate frame.	p. 13
$\dot{x}$, $\ddot{x}$	The time derivative, second time derivative.	p. 23
x^μ or U^μ	Italicized, with a Greek index (e.g., position or particle velocity). The components of a 4-vector: (0, 1, 2, 3).	p. 37

x or **U**	A 4-vector (e.g., position or 4-velocity), as distinct from individual components.	p. 37
$x^0 = t$	The time coordinate in an unbarred frame.	p. 37
$x^1 = x$	The first spatial component of a vector (3- or 4-vector) in an unbarred frame. Note (t, x) are ordinary spacetime components.	p. 37
v	The relative speed of two frames. Used more or less interchangeably with the one-dimensional velocity of a frame in most examples.	p. 43
τ	The proper time between two spacetime events. For timelike separations, $d\tau = ds$.	p. 51
$\partial_\mu = \frac{\partial}{\partial x^\mu}$	The derivative with respect to a coordinate.	p. 123

A.2   Tensor Rules

1. What you label an index doesn't really matter.

   By itself, x^i and x^j mean the same thing: the components of a position 3-vector.

2. Equations need to have matching indices.

   For instance,

$$F^i = ma^i$$

is a valid equation, while

$$F^i = ma^k$$

is not.

   What's more, whether the indices are upstairs or downstairs definitely matters, as does the order. For instance,

$$x^\mu \neq x_\mu$$

and

$$F^{\mu\nu} \neq F^{\nu\mu}.$$

3. The Einstein summation convention.

If you have tensors with the same index upstairs and downstairs, you sum over them automatically. For instance,

$$v^i u_i = v^1 u_1 + v^2 u_2 + v^3 u_3.$$

A couple of quick warnings: (a) $v^i u^i$ is meaningless. The matching indices need to be 1 upstairs, 1 downstairs. (b) The indices need to match. If I wrote down $v^i u_k$, they wouldn't sum. Instead, you'd get a rank-2 tensor.

4. The metric is used to compute the dot products of two vectors.

Using the rules above (and assuming the metric is diagonal—which it is in every example we use),

$$g_{ij} u^i v^j = \vec{u} \cdot \vec{v} = g_{11} u^1 v^1 + g_{12} u^1 v^2 + \cdots + g_{33} u^3 v^3 = \text{scalar}.$$

Or, with 4-vectors,

$$g_{\mu\nu} A^\mu B^\nu = \mathbf{A} \cdot \mathbf{B} = g_{00} A^0 B^0 + g_{01} A^0 B^1 + + g_{02} A^0 B^2 + \cdots + g_{33} A^3 B^3.$$

5. The metric can raise and lower indices.

It can turn vectors into 1-forms, and the inverse can do the opposite. For the Minkowski metric, we have two very closely related versions of the metric:

$$g_{\mu\nu} = \begin{pmatrix} 1 & 0 & 0 & 0 \\ 0 & -1 & 0 & 0 \\ 0 & 0 & -1 & 0 \\ 0 & 0 & 0 & -1 \end{pmatrix} ; \; g^{\mu\nu} = \begin{pmatrix} 1 & 0 & 0 & 0 \\ 0 & -1 & 0 & 0 \\ 0 & 0 & -1 & 0 \\ 0 & 0 & 0 & -1 \end{pmatrix}.$$

Using the summation rule above, the metric lowers

$$U_\beta = g_{\alpha\beta} U^\alpha,$$

or raises

$$p^\mu = g^{\mu\nu} p_\nu,$$

tensor indices.

6. Spacetime derivatives make an additional index downstairs.

You can take a derivative of a scalar, a vector, or any tensor. For instance,

$$\partial_\mu \Phi = \left(\frac{\partial \Phi}{\partial t} \quad \frac{\partial \Phi}{\partial x} \quad \frac{\partial \Phi}{\partial y} \quad \frac{\partial \Phi}{\partial z} \right).$$

A.3   Vector Operations in Cartesian Coordinates

We use a number of operations frequently throughout the text. They are collected here (along with their tensor notation versions) for ease of reference. We also include a few vector relations that are true in general.

Gradient:

$$\nabla U = \partial_i U \vec{e}^{\,i}$$

$$= \frac{\partial U}{\partial x}\hat{i} + \frac{\partial U}{\partial y}\hat{j} + \frac{\partial U}{\partial z}\hat{k}$$

Divergence:

$$\nabla \cdot \vec{v} = \partial_i v^i$$

$$= \frac{\partial v^1}{\partial x} + \frac{\partial v^2}{\partial y} + \frac{\partial v^3}{\partial z}$$

Curl:

$$\nabla \times \vec{A} = \begin{vmatrix} \hat{i} & \hat{j} & \hat{k} \\ \partial_1 & \partial_2 & \partial_3 \\ A^1 & A^2 & A^3 \end{vmatrix}$$

$$= \left(\partial_2 E^3 - \partial_3 E^2\right)\hat{i} + \left(\partial_3 E_1 - \partial_1 E^3\right)\hat{j} + \left(\partial_1 E^2 - \partial_2 E^1\right)\hat{k}$$

Laplacian:

$$\nabla^2 \Phi = -\partial_i \partial^i \Phi$$

$$= \frac{\partial^2 \Phi}{\partial x^2} + \frac{\partial^2 \Phi}{\partial y^2} + \frac{\partial^2 \Phi}{\partial z^2}$$

D'Alembertian

$$\Box \Phi = \partial_\mu \partial^\mu \Phi$$

$$= \frac{\partial^2 \Phi}{\partial t^2} - \frac{\partial^2 \Phi}{\partial x^2} - \frac{\partial^2 \Phi}{\partial y^2} - \frac{\partial^2 \Phi}{\partial z^2}$$

$$= \ddot{\Phi} - \nabla^2 \Phi$$

Divergence Theorem:

$$\int d^3x \nabla \cdot \vec{J} = \oiint dS \vec{J} \cdot \hat{n}$$

Answers to Selected Problems

The answers, by which I mean the final results, to most odd problems appear below. Instructors who stress evaluating students' work should consult the solution key for more detailed derivations. Problems that are solely focused on a derivation or a proof of a result presented elsewhere are not included.

Chapter 1

1. (a) Not valid;
   (b) Valid;
   (c) Not valid;
   (d) Valid

3. (b) $25, 5$;
   (c) -10;
   (d) $\begin{pmatrix} -2\sqrt{3} + \frac{3}{2} \\ 2 + \frac{3\sqrt{3}}{2} \end{pmatrix} \approx \begin{pmatrix} -1.964 \\ 4.598 \end{pmatrix}$;
   (e) $B^{\vec{\imath}} = \begin{pmatrix} -0.134 \\ -2.232 \end{pmatrix}$;
   (g) -10

5. (a) $\begin{pmatrix} 0 & 2 \\ 1 & 0 \end{pmatrix}$;
   (b) $\begin{pmatrix} 0 & 1 \\ 1/2 & 0 \end{pmatrix}$;
   (c) $\begin{pmatrix} 1/4 & 0 \\ 0 & 1 \end{pmatrix}$

Chapter 2

1. (a) 500 s;
   (b) 1.28 s;
   (c) 1.02×10^{-9} s

3. (a) (i) 0; (ii) −3; (iii) 3;
   (b) (i) BC; (ii) AC; (iii) AB;
   (c) 0.5

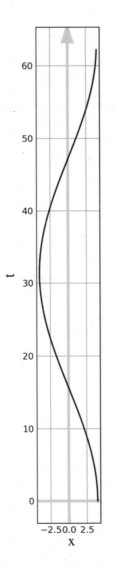

5.

Chapter 3

1. (a) 23 min, 6 sec

3. (a) 0.045;
   (b) 0.6;
   (c) 0.98;
   (d) 9999994

5. (a) -9;
   (b) $\begin{pmatrix} \frac{40}{3} \\ \frac{41}{3} \end{pmatrix}$;
   (c) -9

7. (a) 20 yr
   (b–d, f);

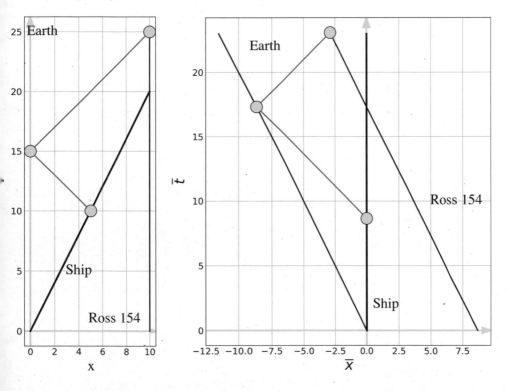

   (e) 17.32 yr

9. (a) $\begin{pmatrix} 5/3 \\ 4/3 \\ 0 \\ 0 \end{pmatrix}$;

(b) 1;

(c) $\begin{pmatrix} 1.705 \\ 1.381 \end{pmatrix}$;

(d) $\begin{pmatrix} 0.039 \\ 0.048 \end{pmatrix}$

Chapter 4

1. 0.999998

3. (a) 0.33 s;

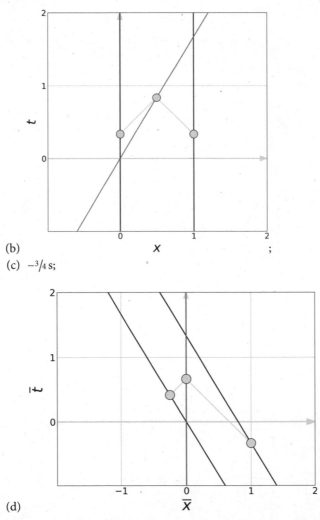

(b)   ;

(c) $-\tfrac{3}{4}$ s;

(d)

5. (a) $1 + \frac{r^2\omega^2}{2}$;

   (b) $r\omega^2, -\frac{r^2\omega^2}{2}$;

   (c) $1 - \Phi$

Chapter 5

1. (a) -10^{-7};

   (c) 0.73;

   (d) -0.423

3. (a) $511,000$ eV;

   (b) 13.8 eV;

   (c) 3800 eV

5. (a) 8×10^{20} J;

   (b) 1.5×10^{21} J;

   (c) 6×10^{20} J;

   (d) 2.88×10^{20} J;

   (e) 1.2×10^{21} j;

   (f) 4×10^{12} kg m/s

7. (a) 5 MeV;

   (b) 5.1 MeV;

   (c) 0.9998

Chapter 6

1. $\frac{Ak}{\omega} \sin(kx - \omega t) + \rho_0$

3. (a) $\dot{\rho} + \frac{\partial(v\rho)}{\partial x} = 0$;

   (b) $\frac{\partial(\rho v)}{\partial t} + \frac{\partial(\rho v^2)}{\partial x} = 0$;

   (c) $\dot{\delta} = -\frac{\partial v}{\partial x}$;

   (d) $\rho_0 \dot{v} = -\frac{\partial P}{\partial x}$

Chapter 7

1. (a) $\gamma^2 \frac{\lambda^2}{2\pi r}$;

   (b) $\frac{\lambda^2 \gamma^2 v^2}{2\pi r}$;

   (c) Same

3. (a) $\begin{pmatrix} 0 & 0 & 0 & E_0 \\ 0 & 0 & 0 & 0 \\ 0 & 0 & 0 & 0 \\ -E_0 & 0 & 0 & 0 \end{pmatrix}$;

(b) $\begin{pmatrix} 0 & 0 & 0 & \gamma E_0 \\ 0 & 0 & 0 & v\gamma E_0 \\ 0 & 0 & 0 & 0 \\ -\gamma E_0 & -v\gamma E_0 & 0 & 0 \end{pmatrix}$;

(c) $\vec{B} = -E_0\gamma\hat{j}$

5. (a) $\begin{pmatrix} \gamma & -v\gamma & 0 & 0 \end{pmatrix}$;

(b) $\begin{pmatrix} 0 & E_0 & 0 & 0 \\ -E_0 & 0 & 0 & -B_0 \\ 0 & 0 & 0 & 0 \\ 0 & B_0 & 0 & 0 \end{pmatrix}$;

(c) $\begin{pmatrix} v\gamma E_0 \\ \gamma E_0 \\ 0 \\ v\gamma B_0 \end{pmatrix}$;

(d) vE_0

Chapter 8

1. (a) -7.4×10^{-10};
   (b) -2.1×10^{-6};
   (c) -0.00021;
   (d) -0.19

3. (a) L/c;
   (b) $\frac{1}{2}g\frac{L^2}{c^2}$;
   (c) $\frac{g^2L^2}{2c^4}$;
   (d) $\frac{\Phi_G}{c^2}$

Bibliography

[1] Barton, John, Breyanna Blackwell, Tina G. Butcher, Richard A. Harshman, and G. Diane Lee, eds. *Specifications, Tolerances, and Other Technical Requirements for Weighing and Measuring Devices*. NIST Handbook 44, 2020 edition. National Institute of Standards and Technology, 2019.

[2] Baudis, Laura, P. A. Zyla, R. M. Barnett, J. Beringer, O. Dahl, D. A. Dwyer, D. E. Groom, et al. "Review of Particle Physics." *Progress of Theoretical and Experimental Physics* 8 (2020): 083C01.

[3] Bell, J. S. "How to Teach Special Relativity." *Progress in Scientific Culture* 1, no. 2 (1976): 1–13.

[4] Carnot, Sadi. *Reflections on the Motive Power of Fire, and on Machines Fitted to Develop That Power*. Bachelier, 1824.

[5] Ehrenfest, Paul. "Uniform Rotation of Rigid Bodies and the Theory of Relativity." *Physicalische Zeitschrift* 10 (1909): 918.

[6] Einstein, Albert. *Autobiographical Notes: A Centennial Edition*. Edited and translated by Paul Arthur Schilpp. Open Court Publishing Co., 1979.

[7] Einstein, Albert. "Does the Inertia of a Body Depend upon Its Energy-Content?" *Annalen der Physik* 18, no. 13 (1905): 639–41.

[8] Einstein, Albert. "On a Heuristic Point of View About the Creation and Conversion of Light." *Annalen der Physik* 17, no. 6 (1905): 132–48.

[9] Einstein, Albert. "On the Electrodynamics of Moving Bodies." *Annalen der Physik* 17, no. 10 (1905): 891–921.

[10] Einstein, Albert. "On the Movement of Small Particles Suspended in Stationary Liquids Required by the Molecular-Kinetic Theory of Heat." *Annalen der Physik* 17 (1905): 549–60.

[11] Einstein, Albert. "On the Relativity Principle and the Conclusions Drawn from It." *Jahrbuch der Radioaktivität und Elektronik* 4 (1907): 411–62.

[12] Einstein, Albert. *Relativity*. Pi Press, 2005. Quotes taken from "The Relativity of Simultaneity," pp. 34–37.

[13] Einstein, Albert, and Hermann Minkowski. *The Principle of Relativity: Original Papers by A. Einstein and H. Minkowski*. University of Calcutta, 1920.

[14] FitzGerald, Edward, and Daniel Karlin, ed. *The Rubáiyát of Omar Khayyám*. Oxford University Press, 2010.

[15] FitzGerald, George Francis. "The Ether and the Earth's Atmosphere." *Science* 13, no. 328 (1889): 390–91.

[16] Fizeau, Hippolyte. "Sur une expérience relative á la vitesse de propagation de la lumiére."*Comptes rendus de l'Académie des sciences* 29 (1849): 90–92, 132.

[17] Foucault, Jean Bernard Léon. "Détermination expérimentale de la vitesse de la lumiére: Parallaxe du soleil." *Comptes rendus de l'Académie des sciences* 55 (1862): 501–3, 792–96.

[18] Gaia Collaboration, S. A. Klioner, F. Mignard, L. Lindegren, U. Bastian, P. J. McMillan, J. Hernández, et al. "Gaia Early Data Release 3: Acceleration of the Solar System from Gaia Astrometry." *Astronomy & Astrophysics* 649 (2021): A9.

[19] Galilei, Galileo. *Dialogue Concerning Two New Sciences*. Macmillan, 1914.

[20] GRAVITY Collaboration, R. Abuter, A. Amorim, M. Bauböck, J. P. Berger, H. Bonnet, W. Brandner, et al. "A Geometric Distance Measurement to the Galactic Center Black Hole with 0.3% Uncertainty." *Astronomy & Astrophysics* 625 (2019): L10.

[21] Heath, Thomas L., and Euclid. *The Thirteen Books of Euclid's Elements, Books 1 and 2*. Dover Publications, 1956.

[22] Huygens, Christiaan. *Treatise on Light*. Pieter van der Aa, 1690. Translation: MacMillan, 1912.

[23] Kapteyn, Jacobus C. "Uber die Bewegung der Nebel in der Umgebung von Nova Persei." *Astronomische Nachrichten* 157, no. 12 (Dec. 1901): 201.

[24] Langevin, Paul. "The Evolution of Space and Time." *Scientia* 10 (1911): 47.

[25] Lorentz, Hendrik. "De relatieve beweging van de aarde en den aether (The Relative Motion of the Earth and the Aether)." *Zittingsverlag Akad. V. Wet.* 1 (1892): 74–79.

[26] Lorentz, Hendrik. "Electromagnetic Phenomena in a System Moving with Any Velocity Less Than That of Light." *Proceedings of the Academy of Sciences of Amsterdam* (1904): 11–34.

[27] Maxwell, James Clerk. "A Dynamical Theory of the Electromagnetic Field." *Philosophical Transactions of the Royal Society of London* 155 (1865): 459–513.

[28] Michelson, Albert A., and Edward W. Morley. "On the Relative Motion of the Earth and the Luminiferous Ether." *American Journal of Science* 34, no. 203 (1887): 333–45.

[29] Minkowski, Hermann. "Raum and Zeit." *Physikalische Zeitschrift* 10 (1909): 75–88.

[30] Newcomb, Simon. "The Solar Parallax." *Nature* 5 (1871): 60–61.

[31] Newton, Isaac. *The Principia*. Prometheus, 1995.

[32] NIST. "A Turning Point for Humanity: Redefining the World's Measurement System." NIST.gov. Updated November 30, 2020. https://www.nist.gov/si-redefinition/turning-point -humanity-redefining-worlds-measurement-system.

[33] Noether, Emmy. "Invariante Variationsprobleme." *Nachrichten von der Gessellshaft der Wissenshaften zu Göttingen, Math-phys. Klasse* (1918): 235–57. Translated by M. A. Tavel. http://arxiv.org/abs/physics/0503066v1.

[34] Penrose, Roger. "The Apparent Shape of a Relativistically Moving Sphere." *Mathematical Proceedings of the Cambridge Philosophical Society* 55, no. 1 (1959): 137–39.

[35] Pierce, Evan. "The Lock and Key Paradox and the Limits of Rigidity in Special Relativity." *American Journal of Physics* 75, no. 7 (2007): 610–14.

[36] Pound, Robert V., and Glen A. Rebka Jr. "Gravitational Red-Shift in Nuclear Resonance." *Physical Review Letters* 3, no. 9 (1959): 439–41.

[37] Rindler, Wolfgang. "Length Contraction Paradox." *American Journal of Physics* 29, no. 6 (1961): 365–66.

[38] Rømer, Ole. "A Demonstration Concerning the Motion of Light, Communicated from Paris, in the Journal des Scavans, and Here Made English." *Philosophical Transactions (1665–1678)* 12 (1677): 893–94.

[39] Rossi, Bruno, and David B. Hall. "Variation of the Rate of Decay of Mesotrons with Momentum." *Physical Review* 59, no. 3 (1941): 223–28.

[40] Shakespeare, William. *Romeo and Juliet.* Dover, 2014.

[41] Taylor, Edwin F., and John Archibald Wheeler. *Spacetime Physics.* W. H. Freeman, 1992.

[42] Terrell, James. "Invisibility of the Lorentz Contraction." *Physical Review* 116 (1959): 1041–45.

[43] Viereck, George Sylvester. "What Life Means to Einstein." *Saturday Evening Post,* October 1929, 17, 110, 113–14, 117.

[44] Voigt, Woldemar. "On Doppler's Principle." *Nachrichten von der Gessellshaft der Wissenshaften zu Göttingen* (1887): 41–51.

[45] Weyl, Hermann. *Symmetry.* Princeton University Press, 1952.

Index